Amanda Hernandes
William Natale
Léon-Etienne Parent

NUTRITIONAL DIAGNOSIS OF GUAVA

Amanda Hernandes
William Natale
Léon-Etienne Parent

NUTRITIONAL DIAGNOSIS OF GUAVA

Nutritional Balance of Guava Trees from Liming and Fertilization through Compositional Data Analysis.

ScienciaScripts

Imprint

Any brand names and product names mentioned in this book are subject to trademark, brand or patent protection and are trademarks or registered trademarks of their respective holders. The use of brand names, product names, common names, trade names, product descriptions etc. even without a particular marking in this work is in no way to be construed to mean that such names may be regarded as unrestricted in respect of trademark and brand protection legislation and could thus be used by anyone.

Cover image: www.ingimage.com

This book is a translation from the original published under ISBN 978-620-2-80649-7.

Publisher:
Sciencia Scripts
is a trademark of
International Book Market Service Ltd., member of OmniScriptum Publishing Group
17 Meldrum Street, Beau Bassin 71504, Mauritius
Printed at: see last page
ISBN: 978-620-3-38031-6

CURRICULAR DATA OF THE AUTHOR

AMANDA HERNANDES - Born in Pitangueiras - SP, on 17 February 1983, she graduated in Agricultural Engineering in July 2007 from the Faculty of Engineering of Ilha Solteira, Paulista State University "Júlio de Mesquita Filho".
- Unesp, Campus of Ilha Solteira, Ilha Solteira - SP. During the undergraduate course, she did several internships, was monitor of two disciplines, participated in several events, was Scientific Initiation scholar (PIBIC/CNPq) for two years, among other activities undertaken. In July 2009, she obtained her Master?s Degree in Agronomy (Soil Science) from the Faculty of Agricultural and Veterinary Sciences of Jaboticabal, Paulista State University "Júlio de Mesquita Filho" - Unesp, Jaboticabal Campus - SP. During the master, he was a CAPES scholar in the first twelve months, and until the end of the course, the CNPq. In August 2009, he started his PhD in Agronomy (Plant Production) in the College of Agricultural and Veterinary Sciences of Jaboticabal, Paulista State University "Júlio de Mesquita Filho" - Unesp, Jaboticabal - SP. During his doctorate, he was a CAPES scholar. From October 2009 to March 2010 and from July 2011 to December 2011, he did his PhD sandwich at *Université Laval*, Quebec in Canada, with the *Departament des Sols et de Génie Agroalimentaire*. He has carried out several activities such as teaching internship, participation in symposia and congresses, participation in research groups, conducting experiments and organizing events. During her academic life, she has published several abstracts in event annals, in addition to scientific articles in journals with editorial board, both national and international. He currently works with Rural Extension at the Secretary of Agriculture and Supply - SAA, of the Government of the State of São Paulo.

*"No one is so great that he
cannot learn, nor so small that he
cannot teach." (Aesop)*

To my beloved parents, Antonio Martins Hernandes Filho and Marta Célia de
Agostini Hernandes, and my brother Antonio Martins Hernandes Neto,
for their unconditional love and for always supporting and encouraging
me in all stages of my life,
I DEDICATE

Contents

I DEDICATE.. 3

CHAPTER 1 - GENERAL CONSIDERATIONS.. 8

CHAPTER 2 - Soil cation balance in guava orchards... 35

CHAPTER 3 - Nutritional balance of guava trees based on fertilisation and liming 88

NUTRITIONAL DIAGNOSIS OF GUAVA

The response of guava to fertilization and liming can be monitored by soil and plant tissue analysis, a tool used to rebalance the soil and the plant by adding nutrients in adequate concentration. However, the concept of base saturation of the cation exchange capacity (CEC), used for the purposes of soil fertility evaluation, is weakly supported experimentally, is intrinsically influenced by the tendency and redundancy of the data, possessing scale dependence and a non-normal distribution. On the other hand, the nutrient profile, used for the diagnosis of plant nutrition, is defined in relation to patterns of nutrient contents, which are constantly criticised for not considering the interactions that occur among the elements and for generating numerical trends, due to data redundancy, scale dependence and non-normal distribution. Thus, the objective of the work was to evaluate a system of isometric log ratios (*ilr*) to correct such distortions, both in the evaluation of soil fertility and in foliar diagnosis, enabling the management of fertilization and liming in guava orchards, adjusting the current nutrient standards with the equilibrium range of the most productive guava trees. For this, a database of data from experiments conducted for three years with doses of N, P and K and, for seven years, with doses of lime, in guava orchards, was used. The soil values of K, Ca, Mg and potential acidity (H + Al) were divided into three binary partitions as follows: the soil balance [K | Ca, Mg, H + Al] to manage K fertilization; the soil balance [Ca, Mg | H + Al] to monitor the need for lime; and, the soil balance [Ca | Mg] to establish liming material. The foliar balances [N, P, K | Ca, Mg], [N, P | K], [N | P] and [Ca | Mg] to separate the effects of fertilizers (N-K) and liming (Ca-Mg) on nutrient balance. The critical soil cation balances were defined without numerical trends, thus being an optimal soil fertility index for fertilizer management in guava orchards. The productivity of the guava trees, as well as their foliar nutrient balance, allowed the definition of nutrient balance ranges, combined in *ilr* coordinates, and their validation with the critical concentration ranges currently used in Brazil.

Keywords: compositional data analysis, soil fertility

soil, Aitchison geometry, plant mineral nutrition, *Psidium guajava* L., isometric log ratios (*ilr*)

GUAVA TREE NUTRITIONAL DIAGNOSIS

SUMMARY - The response of the guava tree to fertilizing and liming can be monitored by analyzing the soil and the plant tissue so that a balance can be re- established by adding nutrients at the appropriate concentration. However, the concept of base saturation of cationic exchange capacity (CEC), used to assess soil fertility, has little experimental backing and is intrinsically influenced by the trend and redundancy of the data, is scale-dependent and has non-normal distribution. On the other hand, the nutritional profile used for plant nutritional diagnosis is defined in relation to standard nutrient content figures which are constantly criticized for ignoring interactions between elements and for generating numeric trends as a result of data redundancy, scale-dependence and non-normal distribution. Therefore, the aim of the study was to evaluate a system of isometric log ratios (*ilr*) to correct these distortions, both in soil fertility assessment and in foliar diagnosis, making it possible to manage fertilizing and liming in guava tree orchards, adjusting the current nutrient standards to the range of the highest yielding guava trees. To do this, a database of experiments conducted in guava tree orchards were used, over a three-year period using doses of N, P and K, and over a seven-year period with doses of lime. The soil values for K, Ca, Mg and potential acidity (H + Al) were split into three binary partitions as follows: soil balance [K | Ca, Mg, H + Al] to manage potassium fertilizing; soil balance [Ca, Mg | H + Al] to monitor the need for liming; and soil balance [Ca | Mg] to establish the liming material. The [N, P, K | Ca, Mg], [N, P | K], [N | P] and [Ca | Mg] foliar balances were selected in order to separate the parts played by fertilizers (N-K) and liming (Ca-Mg) in balancing the nutrients. The critical cationic balances in the soil were defined with no numeric trends, providing an excellent soil fertility index for managing fertilizer application in guava orchards. Guava tree yields and foliar nutritional balance were used to define nutrient balance ranges combined in *ilr* coordinates, and validated against the critical concentration ranges currently used in Brazil.

Keywords: compositional data analysis, soil fertility, Aitchison geometry, plant mineral nutrition, *Psidium guajava* L., isometric log ratios (*ilr*)

CHAPTER 1 - GENERAL CONSIDERATIONS

Introduction

Brazilian fruit farming has been gaining space in the world market, transforming Brazil into a major exporter of fruit and creating business opportunities for farmers in highly profitable ventures. Guava, a highly nutritious fruit whose plant is native to Tropical America, occupies a prominent place among tropical fruits, placing Brazil in the position of the world's largest producer of red guava (IEA, 2012). However, the average production of guava (*Psidium guajava* L.) orchards in the state of São Paulo is considered low, due, among other factors, to inadequate nutrient management because of the lack of information about the nutrition, fertilization and liming of the crop.

Adequate management of liming and fertilization is essential for a balance between the supply of nutrients, aiming to obtain high yields with fruits of excellent quality. Thus, the guava requires substantial quantities of nutrients to reach this high productivity and fruit quality, being responsive to fertilization and liming, a fact confirmed by several field experiments, conducted over several years and in different climates and soils (ARORA; SINGH, 1970; NATALE, 1993; NATALE et al., 1994; NATALE et al., 1996; NATALE, 1999; NATALE et al., 2001; NATALE et al., 2002ab; PRADO, 2003; PRADO; NATALE,
2004, 2008; NATALE et al., 2005; PRADO et al., 2005; NATALE et al., 2007; ANJANEYULU; RAGHUPATHI, 2009; OSMAN; ABD EL-RAHMAN, 2009; ROZANE et al., 2009; SOUZA et al., 2009).

Soil and plant tissue analyses can help to monitor the response of guava to fertilization and liming, but their methods of interpretation use conceptually biased approaches, since they do not take into account the interactions that occur between the elements. Furthermore, the concepts used in the interpretation of the analyses generate numerical trends resulting from

of data redundancy, and have scale dependence and non-normal distribution (PARENT, 2011).

To avoid the distortions that occur in the interpretations of soil and plant tissue analyses, the concept of isometric log ratio transformations (*ilr*), proposed by Egozcue et al. (2003), which is unbiased and non-biased, can be used. Thus, the hypothesis of the present study is that the interpretations of soil and plant tissue analyses using the *ilr* concept are better than the interpretations arising from the use of other approaches. With this, the objective of the present study was to evaluate the feasibility of using this concept in the interpretation of soil fertility and plant nutrition assessment.

REVIEW OF LITERATURE

Tropical fruit farming

Fruit farming has occupied a prominent place in world agriculture, mainly because of the possibility of using soils considered unsuitable for traditional agriculture. Brazil stands out as one of the three largest fruit producers in the world, with a volume of 43 million tons year-1 (IBRAF, 2012), a position achieved due to favourable conditions of climate, soil and availability of area. This sector has been developing rapidly, with positive effects on the national economy. However, the country's productivity is still considered low when compared to that of traditional fruit-producing countries, due to several factors, including the scarcity of research results on production technology and, in particular, on liming and fertilization practices.

Despite being considered a major fruit producer, Brazil has low *per capita* consumption of fresh fruits. According to IBRAF (2012), the national consumption is around 57 kg per person per year, which is considered low when compared to other countries such as Spain (120 kg per year), Germany (112 kg per year), the United States (67 kg per year) and Japan (62 kg per year). It is worth noting that the World Health Organization (WHO) recommends a *per capita* consumption of 146 kg of fruit per year.

The performance of Brazilian fruit farming shows that the country has the potential to produce quality fruit, meeting the most demanding requirements of foreign markets. Brazilian fruits represent only 2% of the international market, which moves US$ 21 billion per year and the country occupies the twentieth place among exporters (BUAINAIN; BATALHA, 2007). It is worth noting that neither the domestic nor the international market is fully supplied with fruit, and both should continue to grow. The world fruit market is currently growing at an average of US$ 1 billion per year (FAO, 2012).

Brazil is in a privileged position in relation to other nations, due to its extensive area and climatic diversity, enabling cultivation in its territory, both of temperate climate plants in the South and Southeast, and of tropical and subtropical climate plants in the Southeast, North and Northeast of the country.

Fruit farming in the state of São Paulo occupies a prominent position in Brazil, both in terms of cultivated area and volume produced, with orchards covering more than 800,000 hectares in the state and 19 million tons of fruit produced annually, representing 36% of the total area of the country occupied by fruit farming and 44% of total Brazilian fruit production (IBGE, 2012). The state of São Paulo offers advantageous conditions in terms of location, soils, climate, infrastructure, agricultural standards and production on an economic basis.

General aspects of the guava tree

The guava tree (*Psidium guajava* L.) is native to the tropical region of the Americas, located between Mexico and Brazil. Currently, this species is distributed worldwide, due to its ability to adapt to different soil and climatic conditions (KOLLER, 1979; GONZAGA NETO, 2001).

A member of the *Myrtaceae* family, the guava starts producing between 10 and 12 months after planting, reaching maximum productive efficiency at the age of 3 or 4 years (ROVIRA, 1988). It is considered a hardy plant, adapting to various types of soil, but deep, permeable soils rich in organic matter provide the best economic results (KOLLER, 1979), provided they also adequately meet the nutritional requirements of the plant, which involves the management of liming and fertilization.

The reproduction of the native guava is via seeds, but it can also be propagated asexually by grafting, budding or pricking out. However, the use of herbaceous cuttings is recommended for the formation of commercial orchards, using improved cultivars that are highly productive but demanding in terms of nutrition (NATALE et al., 1996). This procedure is routinely used by nurserymen to produce guava seedlings (TODA

FRUTA, 2012), due to the genetic uniformity for the formation of orchards (PEREIRA; NACHITIGAL, 1997).

Among the Brazilian tropical fruits, the guava occupies a prominent place, not only for its aroma and flavor, but also for its nutritional value and high content of vitamin C, besides providing carbohydrates, protein, fiber and vitamins (MANICA et al., 2000), placing Brazil in the position of world's largest producer of red guava (IEA, 2012). Its fruits are used not only in the industry, *in* multiple forms (puree, pulp, nectar, juice, compote, ice cream, among others), but are also widely consumed *fresh* (GONZAGA NETO, 2001).

In 2007, Brazil produced more than 300,000 tons of guava in an area exceeding 15,000 hectares, but only 0.07% of this production was destined for export (IBGE, 2012), with the European Union as the main destination, which today concentrates 70% of the volume of Brazilian exports. This highlights the need to improve the quality of the fruits in order to gain space in the international market.

Guava culture has evolved greatly in recent years, mainly due to the development of more productive cultivars, which require, however, better cultural treatments, requiring more intensive management. Thus, systems that serve both the *fresh* and industrial markets require adequate management, such as pruning and irrigation (SERRANO et al., 2007), and the application of correctives and fertilizers (NATALE et al., 1996).

There is no doubt that there are possibilities of increasing the participation of guava in Brazilian exports, but for that it is essential that there is greater rationalization of its cultivation, from planting, to the essential care with the aspects inherent to the presentation and regularity of the product supply in the international market (GONGATTI NETTO et al., 1994).

In Brazil, the largest guava-growing regions are the Southeast and Northeast, with São Paulo and Pernambuco being the largest producers (IBGE, 2012). In São Paulo, the municipalities of Monte Alto, Taquaritinga, Vista Alegre do Alto and Urupês account for 35% of the country's guava production (IBGE, 2012). This region has the advantage of having agro-industries located in the municipalities of Matão, Monte Alto, Taquaritinga and Vista Alegre do Alto (FRANCISCO et al., 2012).

The average production of orchards in São Paulo is about 14 t ha-1 for table and 25 t ha-1 for industry (SÉRIE..., 2001), which is clearly low productivity, attributed, among other factors, to the lack of information about nutrition, fertilization and liming of the crop (NATALE et al., 1996).

Mineral nutrition

Plants, in general, need essential elements for adequate growth and development, such as C, H, O, from the air and water, which make up approximately 95% of the dry mass of the plant. They also need macronutrients (N, P, K, Ca, Mg and S), micronutrients (Fe, Mn, Zn, Cu, B, Cl and Mo) (RAIJ, 1991) and beneficial or essential elements (Se, Co, Ni, Si) for certain groups of plants (MALAVOLTA, 2006).

The productivity and quality of the fruit in an orchard is the result of the interaction of several factors, especially the genetic potential, as well as the management of the soil, nutrients and the water balance. For maximum production and optimum quality of the fruit, it is necessary to have a balanced supply of nutrients.

Despite the significant production of guava, there has been little research into the mineral nutrition of this fruit. In general, in the literature, the guava plant has been considered rustic, tolerant to acidity (GUERRERO, 1991) and not very demanding in terms of soil (PEREIRA; MARTINEZ JUNIOR, 1986). According to Clarkson (1985), this fruit has a high capacity and/or efficiency of nutrient use, when cultivated in soils with low fertility. Thus, it is to be expected that the plant has a low capacity to respond to the application of nutrients, or the correction of soil acidity. However, Nachtigal et al. (1994) observed that the mountain guava, a fruit of the *Myrtaceae* family, originating in acidic soils with low fertility, showed a positive response when subjected to soil fertility improvement. Therefore, to obtain satisfactory economic results in commercial orchards it is necessary to maintain soil fertility at adequate levels, promoting substantial increases in fruit production (NATALE, 1993).

Mineral nutrition represents one of the most important aspects for fruit trees, as for the majority of plants. The nutritional requirements of guava are relatively high, but the poverty of the soils where they are grown makes the application of almost all of the mineral elements essential for the full development of the plants mandatory (NATALE, 1999).

Nitrogen is part of amino acids, proteins and nucleic acids, and plays an important role in processes like ionic absorption, photosynthesis, respiration, cell division and differentiation (MALAVOLTA et al., 1989). Thus, it is fundamental to the structure of the plant, facilitating growth, flowering and fruit formation and, consequently, the productivity of the guava plant.

Phosphorus is directly related to the structural functions and processes of energy transfer/storage, and also plays a fundamental role in the process of photosynthesis and cell multiplication and division. Thus, the development of fruit trees at the moment of their establishment is favoured by sources of P available in the soil, which stimulate root growth and, as a reflex, the development of the aerial part (MARSCHNER, 1995). Therefore, low availability of P in the soil may markedly reduce the growth of guava seedlings (RODRIGUEZ et al., 1968; SALVADOR et al., 1998).

Potassium acts in the cell turgor mechanism, in the transport of carbohydrates and in the quality of the fruit (MARSCHNER, 1995). Although K is not a component of any organic plant compound, it is important in the transport and assimilation of carbohydrates, which are also necessary for the formation of cell walls. Thus, K promotes, by the transport of photo-assimilates, an increase in the weight of the fruit, and improvements in the taste and storage capacity of the guava fruits. Thus, when there is a deficit in the supply of K, a reduction in the growth of the shoot of the guava tree is to be expected (RODRIGUEZ et al., 1968), since there is a reduction in the plant metabolism.

In general, taking into consideration some works developed with the guava, it can be affirmed that the requirements of the fruit in N and K are higher than those in P (BRASIL SOBRINHO et al., 1961;. HIROCE et al., 1977; KHANDUJA; GARG, 1980; MEDINA, 1988; QUEIROZ et al., 1986; NATALE, 1993; NATALE, 1999), which does not mean that the plant does not need this nutrient for suitable development. Fertilizing fruit trees should also consider the difficulty in combining the quality of the harvested product with productivity, because the nutritional aspect may affect important characteristics of the fruit, such as colour, flavour and size, among others (MALAVOLTA, 1994). Thus, the increase in production, at the expense of increasing the amount of fertilizer, may cause a reduction in the quality of the fruit, affecting size, resistance to transport and storage, internal and external coloration and the content of soluble solids (MALAVOLTA, 1989).

Excessive concentrations of one element in the soil may lead to a reduction in the absorption or even hinder the use of another element. Thus, it is necessary to study the interaction of nutrients, from the use of production data to studies on the biochemical behaviour in plants (NATALE, 1993).

Liming

It is known that the majority of Brazilian soils are acidic and with low fertility, due to the original material and the intense process of weathering that they have undergone. These soils are characterized by low concentrations of Ca and Mg and high quantities of Al and Mn, as well as less availability of other nutrients and greater adsorption of P. These factors limit the adequate growth and development of the guava plant.

Several factors may be responsible for low crop productivity in tropical areas, especially the limited capacity of soils to meet the nutritional requirements of plants and inadequate management of liming and fertilization (NATALE, 2009). Thus, one of the determinants of increased crop productivity is meeting the nutritional requirements of plants, especially through liming and fertilization.

In addition, as a result of genetic improvement, the plants started to produce more and with higher quality, but their demand for nutrients also

umentou. However, Brazilian soils are naturally poor in terms of fertility and are subjected to constant exploitation, leading them to exhaustion. Therefore, liming and fertilization are imperatives for sustainable farming and soil conservation as a natural resource (NATALE, 2009).

Fruit orchards are long-term agricultural holdings, whose plant roots remain practically restricted to the same volume of soil for several years. Moreover, soil acidity is recognized as one of the main factors of low crop productivity (RAIJ, 1991). Thus, the homogeneous incorporation of lime in depth will provide an adequate environment for root growth, which will permit the efficient use of water and nutrients contained in the corrected layers, positively affecting the development and nutritional state of the fruit tree, resulting in the rational use of fertilizers and improving the benefit/cost ratio, by increasing productivity (NATALE, 2009).

The importance of the plant's root system is obvious, since there is a close dependence between the development of the roots and the formation of the aerial part. The greater or lesser success of lime application depends, in turn, on the nature of the root system and the volume of soil effectively exploited by the crop. Thus, correcting the acidity of the soil is the most efficient way to eliminate chemical barriers to the full development of the roots and, consequently, of the plant (PRADO, 2003).

In acidic soils, with high aluminium saturation, liming promotes the precipitation of Al from the superficial layers, making intense root proliferation possible, with positive effects on plant growth. Furthermore, liming provides considerable quantities of calcium and magnesium, which are macronutrients fundamental for the development of the guava plant.

One of the main functions of calcium is structural, as an integrant of the cell wall, interfering in the formation of pectin and in the organization of the middle lamella, which can influence the texture, firmness and maturation of the guava fruits (MENGEL; KIRKBY, 2000). It is also an enzyme activator, influencing and favoring various metabolic processes in the fruit, thus contributing to the adequate development of the guava plant.

agnesium's main functions in plants include enzymatic and structural activator, participating in the formation of chlorophyll. The low availability of magnesium therefore causes a decrease in the synthesis of chlorophyll and consequently in the photosynthetic rate. Thus, with the decrease in the synthesis of carbohydrates in the leaves, there is a marked decrease in the growth of the guava plant (RODRIGUEZ, et al., 1968).

Liming also promotes an increase in the levels of soluble solids in the fruit, thus making it possible to reach the point of harvest earlier (PARO et al., 1994). Therefore, the application of lime is fundamental to obtain high productivity and quality.

Thus, knowing the limiting factors for guava production allows the adoption of more rational programmes of liming and fertilization, aiming at improving the nutritional state and, consequently, productivity, with benefits throughout the agricultural chain (NATALE, 2009).

Nutritional diagnosis

I.   **Soil fertility assessment**

Chemical soil analysis is the oldest and most traditionally used method for the evaluation of soil fertility and, through it, the need for correction and fertilization of crops can be established. Its interpretation is based solely on the formation and maintenance of sufficient levels of available nutrients (McLEAN, 1984). However, it is difficult to correctly establish the adequate dose of nutrients, since the soil is a complex, heterogeneous medium, where numerous chemical, physical and microbiological reactions occur, and where nutrients interact, in addition to the complex interactions that occur between the climate, nutrient sources, soils and plants, which may influence the availability and use of the nutrients added with the application of fertilizers. Despite the multitude of interactions

nutrients in the soil (MALAVOLTA, 2006), the observation of the balance between the elements is not part of the soil analysis procedure.

The most common concepts used in the interpretation of soil analysis are nutrient intensity and balance and cationic base saturation rate. The concept of nutrient intensity and balance, much applied to horticultural crops, uses soluble nutrients and total salt concentrations for diagnostic purposes (GERALDSON, 1957; 1970; 1984). The concept of cationic base saturation rate, on the other hand, used more for agronomic crops, is an approach to the balance of exchangeable cations and potential acidity, which interact with each other in the cation exchange capacity (CEC), expressed by double relations, such as K/Ca, K/Mg and Ca/Mg (McLEAN et al, 1983; McLEAN, 1984), although there is no experimental evidence as to what the "ideal" soil cationic relations would be (LIEBHARDT, 1981; KOPITTKE; MENZIES, 2007).

The database arising from these approaches is compositional, that is, the data are expressed as percentages. So, consequently, they become conceptually biased approaches, since they are influenced by scale dependence, redundancy and non-normal distribution of data (PARENT et al., 2011).

K-Ca-Mg relationships can be illustrated by ternary diagrams, where one component can be omitted and calculated by the difference between 100% and the sum of the other two. As the third component will always be redundant, there are then only two degrees of freedom to diagnose a K-Ca-Mg system. If more components are added, such as, for example, potential acidity or sodium (Na), the diagnosis must be carried out using D-1 degrees of freedom, avoiding this redundancy (EGOZCUE; PAWLOWSKY-GLAHN, 2005).

Thus, any proportion that affects another, in case of change, generates at least a spurious correlation. Some authors had already observed and reported the problem of spurious correlations between component proportions and proportions relative to a measurement scale, for example, dry matter, fresh matter or organic matter basis (PEARSON, 1897; TANNER, 1949; CHAYES, 1960). Thus, spurious correlations will distort linear statistical models, since

affect the covariance matrix, and may even produce contradictory results (PARENT et al., 2011).

In contrast to real data, which are free to move in real space $\pm \infty$, compositional data exhibit non-normal distribution, within their closed space between 0 and 100%, since the lower and upper bounds of the confidence interval cannot be negative or greater than 100% (DIAZ-ZORITA et al., 2002; WELTJE, 2002). However, log ratios can assume negative or positive values, varying freely in the real space $\pm \infty$, thus their transformation is important to maintain the ratios within the closed space.

Several transformations can be used for the compositional data, as the additive log ratio transformation (*alr*), the centralized log ratio (*clr*) (AITCHISON, 1986) and the isometric log ratio (*ilr*) (EGOZCUE et al., 2003). The centralized log ratio has been used in studies on plant nutrition (PARENT; DAFIR, 1992; PARENT et al., 2009), soil solution (PARENT et al., 1997) and hydroponics (LOPEZ et al., 2002). The isometric log ratio transformations (*ilr*) are based on D-1 balances organized orthogonally, being then an unbiased and unbiased approach, (D-1 degrees of freedom), preserving all the information contained in the compositional vector, thanks to the orthogonality principle. The *ilr* concept has been successfully tested in geosciences (BUCCIANTI, 2011), as well as in studies of plant nutrition (PARENT, 2011), organic waste decomposition (PARENT et al., 2011) and soil aggregation (PARENT et al., 2012).

II.   Evaluation of plant mineral nutrition

Leaf analysis is the most appropriate criteria for determining the nutritional state of perennial plants, when compared with soil analysis (CHADHA et al., 1973), making it one of the most important tools in research into the mineral nutrition of fruit plants, not only for determining the response to the nutrients applied or confirming the symptoms of deficiency, but also for serving as an auxiliary criterion in fertilizer recommendations. It also has an advantage over soil analysis as a diagnostic tool, because perennial crops, like guava, because they have a deep root system, have access to more nutrients in the soil at depth than would be found through normal soil analysis procedures (SMITH et al., 1985).

The fundamental idea behind the analysis of plant tissues is that the quantity of nutrients absorbed by the plant reflects the quantity of these nutrients available in the soil (LUNDEGÅRDH, 1943), with changes in the supply of nutrients being reflected in the levels of these elements in the plant tissue (BOULD, 1968). One of the fundamentals of foliar diagnosis is based on the premise that it is in the leaves that most of the physiological and metabolic processes take place and that the mineral content of the leaves should always be related to the development and increase of production (SALVADOR et al., 1999). To meet the requirements of production and quality, the nutrients should be contained in the leaves, not only in an adequate concentration, but also in a balanced relationship (MALAVOLTA et al., 1997). Plant tissue analysis and its respective methods of interpretation, normally used to diagnose the nutritional state of plants, are specific for each crop, as well as for the tissue analyzed and the growth phase of the plant (BATES, 1971; BEAUFILS, 1973; PARENT; DAFIR, 1992), and the nutrient contents may be affected by the age of the leaf, crop load, season and sample size. Thus, the diagnosis of nutrients in plant tissues requires standardized methods of sampling and routine analysis, standard nutrient values and an impartial standardized interpretation of analytical results for making fertilization recommendations (KENWORTHY, 1983). However, the standard nutrient contents were criticised for not considering interactions between elements (BATES, 1971), since several double and multiple interactions have been documented in plants (BERGMANN, 1988; MALAVOLTA, 2006).

The concepts of Critical Nutrient Level (BATES, 1971), Integrated Diagnosis and Recommendation System (DRIS) (BEAUFILS, 1973) and Nutrient Composition Diagnosis (NDC) (PARENT; DAFIR, 1992) are the most widely used concepts in the interpretation of plant tissue analysis.

In the concept of the critical nutrient level, the content of an element in the plant tissue should be greater than a certain minimum or critical content for maximum growth and development of the crop. However, the critical content of the elements declines during plant growth due to the dilution effect (BÉLANGER et al., 2001). Relationships between nutrient content and plant dry matter are, therefore, influenced by the crop growth season (BÉLANGER et al., 2001). However, the critical level of nutrients does not consider the fact that leaf composition constitutes a closed information space, delimited only to the measurement unit, where all the nutrients interact.

The DRIS uses double nutrient ratios, and can explain, in part, the interactions between elements (WALWORTH; SUMNER, 1988); however, it is geometrically deficient, as described in Parent (2011). The CND uses the centralized log ratio transformation (PARENT; DAFIR, 1992), used to analyze compositional data, such as nutrient contents (AITCHISON, 1986), to try to correct and adjust the DRIS procedure.

Interaction between nutrients occurs when one element impedes or interferes with the absorption of another (BATES, 1971). Nutrients may also be accumulated as luxury consumption, or at excessive levels, to the detriment of other components. As the nutrients absorbed by the plant interact within the physical limits of the plant tissue, they may generate spurious correlations. The logarithmic transformation proposed by Aitchison (1986) for compositional analysis, used by Parent and Dafir (1992) and Parent et al. (2009) for diagnostic purposes, can reduce spurious correlations among plant tissue components.

The isometric log ratio transformation (*ilr*) (EGOZCUE; PAWLOWSKI- GLAHN, 2005) is also one of the logarithmic transformations that can avoid the

endency and the redundancy inherent in compositional data, reducing spurious correlations between plant tissue components. *Ilr* coordinates can describe nutrient interactions along orthonormal axes, as can nutrient balance as a distance of nutrient norms on the same axes. Thus, tissue analysis and interpretation methods can provide a diagnosis of plant nutrient balance across different growth stages.

References

AITCHISON, J. **The statistical analysis of compositional data.**
London: Chapman & Hall, 1986. 416 p.

ANJANEYULU, K.; RAGHUPATHI, H. B. Identification of yield-limiting nutrients through DRIS leaf nutrient norms and indices in guava (*Psidium guajava*). **Indian Journal of Agricultural Sciences**, New Delhi, v. 79, p. 418-421, 2009.

ARORA, J. S.; SINGH, J. R. Effect of nitrogen, phosphorus and potassium sprays on guava (*Psidium guajava* L.). **Journal of the Japanese Society for Horticultural Science**, Kyoto, v. 39, p. 55-62, 1970.

BATES, T. E. Factors affecting critical nutrient concentrations in plants and their evaluation: a review. **Soil Science**, Baltimore, v. 112, p. 116-130, 1971.

BEAUFILS, E. R. **Diagnosis and recommendation integrated system (DRIS).**
Pietermaritzburg: University of Natal, 1973. 132 p.

BÉLANGER, G.; WALSH, J. R.; RICHARDS, J. E.; MILBURN, P. H.; ZIADI, N. Critical
nitrogen curve and nitrogen nutrition index for potato in Eastern Canada. **American Journal of Potato Research**, Orono, v. 78, p. 355-364, 2001.

BERGMANN, W. **ErnährungsstörungenbeiKulturpflanzen.** [2nd]. ed.
Stuttgart: Auflage, Gustav Fischer Verlag, 1988. 762 p.

BOULD, C. Leaf analysis as a diagnostic method and advisory aid in crop production.
Experimental Agriculture, Dundee, v. 4, p. 17-27, 1968.

BRASIL SOBRINHO, M. O. C.; MELLO, F. A. F.; HAAG, H. P.; LEME JR., J. A chemical composition of guava (*Psidium guajava* L.). **Anais da Escola Superior de Agricultura "Luiz de Queiroz"**, Piracicaba, v. 18, p. 183-192, 1961.

BUAINAIN, A. M.; BATALHA, M. O. **Cadeia produtiva de frutas.** Ministério da Agricultura, Pecuária e Abastecimento, Secretaria de Política Agrícola, Interamerican Institute for Cooperation on Agriculture. Brasília: IICA, MAPA/SPA, 2007. 102 p.

BUCCIANTI, A. Natural laws governing the distribution of the elements in geochemistry: the role of the log-ratio approach. In: PAWLOWSKY-GLAHN, V.; BUCCIANTI, A. (Eds). **Compositional data analysis:** theory and applications. New York: John Wiley and Sons, 2011. p. 255-266.

CHADHA, K. L.; ARORA, J. S.; RAVEL, P.; SHIKHAMANY, S. D. Variation in the mineral composition of the leaves of guava (Psidium guajava L.) as affected by leaf position, season and sample size. **Indian Journal of Agricultural Science**, New Delhi, v. 43, p. 555-561, 1973.

CHAYES, F. On correlation between variables of constant sum. **Journal of Geophysical Research**, Washington, v. 65, p. 4185-4193, 1960.

CLARKSON, D. T. Factors affecting mineral nutrients acquisition by plants. **Annual Review of Plant Physiology**, Palo Alto, v. 36, p. 77-115, 1985.

DIAZ-ZORITA, M.; PERFECT, E.; GROVE, J. H. Disruptive methods for assessing soil structure. **Soil and Tillage Research**, Amsterdam, v. 64, p. 3-22, 2002.

EGOZCUE, J. J.; PAWLOWSKY-GLAHN, V. Groups of parts and their balances in compositional data analysis. **Mathematical Geology**, New York, v.37, p. 795-828, 2005.

EGOZCUE, J. J.; PAWLOWSKY-GLAHN, V.; MATEU-FIGUERAS, G.; BARCELÓ-
VIDAL, C. Isometric log-ratio transformations for compositional data analysis.
Mathematical Geology, New York, v. 35, p. 279-300, 2003.

FAO - Food and Agriculture Organization of the United Nations. **Production.**
Available at: < http://www.fao.org/>. Accessed 29 Apr. 2012.

FRANCISCO, V. L. F. S.; BAPTISTELLA, C. S. L.; AMARO, A. A. **The guava culture
in the State of São Paulo.** Available at: <http://www.iea.sp.gov.br>.
Accessed on: 15 Jun. 2012.

GERALDSON, C. M. Factors affecting calcium nutrition of celery, tomato and pepper.
Soil Science Society of America Proceedings, Madison, v. 21, p. 621-625, 1957.

GERALDSON, C. M. Intensity and balance concept as an approach to optimal vegetable production. **Communications in Soil Science and Plant Analysis**, New York, v. 1, p. 187-196, 1970.

GERALDSON, C. M. Nutrient intensity and balance. In: STELLY, M. (Ed.). **Soil testing:** correlating and interpreting analytical results. Madison: American Society of Agronomy, 1984. p. 75-84.

GONGATTI NETO, G. A.; GARCIA, A. E.; ARDITO, E. F. G.; GARCIA, E. C.; BLEINROTH, E. W.; MATALLO, M.; CHITARRA, M. I. F.; BORADIM, M. R. **Guava
para exportação:** procedimentos de colheita e pós-colheita. Brasília: Embrapa - SPI, 1994. 35 p.

GONZAGA NETO, L. **Frutas do Brasil**: guava - production. Brasília: Embrapa, 2001. 72 p.

GUERRERO, R. Soil acidity - its nature, implications and management. In: MOJICA, F. S. (Ed.). **Fundamentos para la**

interpretación de análises de suelos, plantas y águas para riego.
Bogota: SCCS, 1991. p. 141-163.

HIROCE, R.; CARVALHO, A. M. de; BATAGLIA, O. C.; FURLANI, P. R.;
FURLANI, A.
M. C.; SANTOS, R. R. dos; PARIQUERA, E. E. de; GALLO, J. R.
Composição mineral de frutas tropicais na colheita. **Bragantia**,
Campinas, v. 36, p. 155-164, 1977.

IBGE- **Brazilian Institute of Geography and Statistics**   .
Available at
< http://www.sidra.Ibge.gov.br/cgi-bin/prtabl>. Accessed on: 19 Apr. 2012.

IBRAF - Brazilian Fruit Institute. **Fruticultura:** síntese. Available
at:<http://www.ibraf.org.br/x-es/f-esta.html>. Accessed on: 13 mar.
2012.

IEA - Institute of Agricultural Economics. **A cultura da guava em
São   Paulo.**   2005.   Available   at: <
http://www.iea.sp.gov.br/out/verTexto.php?codTexto=1902>.
Accessed on: 05 feb. 2012.

KENWORTHY, A. L. Leaf analysis as an aid in fertilizing orchards. In:
WALSH, L. M.; BEATON, J. D. (Eds.). **Soil testing and plant
analysis.** Revised edition, 5th printing. Madison: Soil Science Society
of America, 1983. p. 381-392

KHANDUJA, S. D.; GARG, V. K. Nutritional status of guava (*Psidium
guajava* L.) trees in North India. **Journal of Horticultural Science**,
Ashford, v. 55, p. 433-435, 1980.

KOLLER, O. M. **Cultura da goiabeira.** Porto Alegre: Agropecuária, 1979.
44 p.

KOPITTKE, P. M.; MENZIES, N. W. A review of the use of the basic
cation saturation ratio and the "ideal" soil. **Soil Science Society of
America Journal**, Wiscosin, v. 71, p. 259-265, 2007.

LIEBHARDT, W. C. The basic cation saturation ratio concept and lime and potassium recommendations on Delaware's Coastal Plain soils. **Soil Science Society of America Journal**, Madison, v. 45, p. 544-549, 1981.

LOPEZ, J.; PARENT, L. E.; TREMBLAY, N.; GOSSELIN, A. Sulfate accumulation and Ca balance in hydroponic tomato culture. **Journal of Plant Nutrition**, Athens, v. 25, p. 1585-1597, 2002.

LUNDEGÅRDH, H. Leaf analysis as a guide for soil fertility. **Nature**, London, v. 151, p. 310-311, 1943.

MALAVOLTA, E. **ABC da Adubação.** 5ª. ed. São Paulo: Agronômica Ceres Ltda, 1989. 292 p.

MALAVOLTA, E. Leaf analysis in Brazil - present and perspectives. In: VII INTERNATIONAL CITRUS CONGRESS, 2., 1994, Acireale. **Anais...** Acireale, Itália, 1994, p. 570-574.

MALAVOLTA, E. **Manual de nutrição de plantas.** São Paulo, Brazil: Ceres, 2006. 638 p.

MALAVOLTA, E.; VITTI, G. C.; OLIVEIRA, S. A. **Avaliação do estado nutricional das plantas:** princípios e aplicações. 2.ed. Piracicaba: POTAFOS, 1997. 319 p.

MALAVOLTA, E.; VITTI, G. C.; OLIVEIRA, S. A. de. **Avaliação do estado nutricional das plantas:** princípios e aplicações. Piracicaba: POTAFOS, 1989. 201 p.

MANICA, I.; ICUMA, I.; JUNQUEIRA, N. T. V.; SALVADOR, J. O.; MOREIRA, A.;
MALAVOLTA, E. **Fruticultura tropical:** guava. Porto Alegre: Cinco Continentes, 2000. 374 p.

MARSCHNER, H. **Mineral nutrition of higher plants.** 2. ed. London: Academic Press, 1995. 889 p.

McLEAN, E. O. Contrasting concepts in soil test interpretation: sufficiency levels of available nutrients versus basic cation saturation ratios. In: STELLY, M. (Ed.). **Soil testing:** correlating and interpreting the analytical results. Madison: American Society of Agronomy, 1984. p. 39-54.

McLEAN, E. O.; HARTWIG, R. C.; ECKERT, D. J.; TRIPLETT, G. B. Basic cation
saturation ratios as a basis for fertilizing and liming agronomic crops. II. Field studies.
Agronomy Journal, Madison, v. 75, p. 635-639, 1983.

MEDINA, J. C.; Guava - culture. In: ITAL. **Guava:** culture, raw material, processing and economic aspects. 2.ed. Campinas: ITAL; Instituto Campineiro de Ensino Agrícola, 1988. p. 1-120.

MENGEL, K.; KIRKBY, E. A. **Principios de nutrición vegetal.** Basel, Switzerland: International Potash Institute, 2000. 692 p.

NACHTIGAL, J. C.; KLUGE, R. A.; ROSSAL, P. A. L.; VAHL, L. C.; HOFFMANN, A.
Efeito do fósforo no desenvolvimento inicial de mudas de goiabeira serrana. **Scientia Agrícola**, Piracicaba, v. 51, p. 279-283, 1994.

NATALE, W. Adubação, nutrição e calagem na guavaeira. In: NATALE, W.; ROZANE,
D. E.; SOUZA, H. A.; AMORIM, D. A. **Cultura da guava**. 2. vol. Jaboticabal: FCAV, Capes, CNPq, FAPESP, Fundunesp, SBF, 2009. 284 p.

NATALE, W. **Diagnose da nutrição nitrogenada e potássica em duas cultivares de goiabeira (*Psidium guajava* L.), durante três anos.** 1993. 150 f. Thesis (PhD in Soil and Plant Nutrition) - Escola Superior de Agricultura "Luiz de Queiroz", Universidade de São Paulo, Piracicaba, 1993.

NATALE, W. **Resposta da goiabeira à adubação fosfatada.** 1999. 132 p. Thesis - Faculdade de Ciências Agrárias e Veterinárias - Unesp, Jaboticabal, 1999.

NATALE, W.; COUTINHO, E. L. M.; BOARETTO, A. E. PEREIRA, F. M. Effect of
potassium fertilization in 'Rica' guava cultivation. **Indian Journal of Agricultural Sciences**, New Delhi, v. 66, p. 201-207, 1996.

ATALE, W.; COUTINHO, E. L. M.; BOARETTO, A. E., BANZATTO, D. A. Phosphorus
foliar fertilization in guava trees. In: TAGLIAVINI, M. (Ed.). Proceedings ISHS on Foliar Nutrition, **Acta Horticulturae**, Bologna, v. 594, p. 171-177, 2002a.

NATALE, W.; COUTINHO, E. L. M.; BOARETTO, A. E.; BANZATTO, D. A. Nutrient
foliar content for high productivity cultivars of guava in Brazil. In: TAGLIAVINI, M. (Ed.). Proceedings ISHS on Foliar Nutrition, **Acta Horticulturae**, Bologna, v. 594, p. 383- 386, 2002b.

NATALE, W.; COUTINHO, E. L. M.; BOARETTO, A. E.; CENTURION, J. F. Response
of guava (*Psidium guajava* L.) cv. Paluma in formation to phosphate fertilization.
Revista Brasileira de Fruticultura, Jaboticabal, v. 23, p. 92-96, 2001.

NATALE, W.; COUTINHO, E. L. M.; BOARETTO, A. E.; PEREIRA, F. M. La fertilisation
nitrogen from the guava tree. **Fruits**, Paris, v. 49, pp. 205-210, 1994.

NATALE, W.; PRADO, R. M.; MÔRO, F. V. Alterações anatômicas induzidas pelo calcio na parede celular de frutos de goiabeira. **Pesquisa Agropecuária Brasileira**, Brasília, v. 40, p. 1239-1242, 2005.

NATALE, W.; PRADO, R. M.; ROZANE, D. E.; ROMUALDO, L. M. Efeitos da calagem
on soil fertility, nutrition and productivity of guava. **Revista Brasileira de Ciência do Solo**, Viçosa, MG, v. 31, p. 1475-1485, 2007.

OSMAN, S. M.; ABD EL-RAHMAN, A. E. M. Effect of slow release nitrogen fertilization on growth and fruiting of guava under Mid Sinai conditions. **Australian Journal of Basic and Applied Science**, Pakistan, v. 3, p. 4366-4375, 2009.

PARENT, L. E. Diagnosis of the nutrient compositional space of fruit crops. **Revista Brasileira de Fruticultura**, Jaboticabal, v. 33, p. 321-334, 2011.

PARENT, L. E.; CISSÉ, S.; TREMBLAY, N.; BÉLAIR, G. Row-centred log ratios as nutrient indexes for saturated extracts of organic soils. **Canadian Journal of Soil Science**, Ottawa, v. 77, p. 571-578, 1997.

PARENT, L. E.; DAFIR, M. A theoretical concept of compositional nutrient diagnosis. **Journal of the American Society for Horticultural Science**, Mount Vernon, v.117, p.239-242, 1992.

PARENT, L. E.; de ALMEIDA, C. X.; PARENT, S. E.; HERNANDES, A.; EGOZCUE,
J. J.; KÄTTERER, T.; GÜLSER, C.; BOLINDER, M. A.; ANDRÉN, O.; ANCTIL, F.; CENTURION, J. F.; NATALE, W. Compositional analysis for an unbiased measure of soil aggregation. **Geoderma**, Amsterdam, v. 179, p. 123-131, 2012.

PARENT, L. E.; NATALE, W.; ZIADI, N. Compositional nutrient diagnosis of corn using the Mahalanobis distance as nutrient imbalance index. **Canadian Journal of Soil Science**, Ottawa, v. 89, p. 383-390, 2009.

PARENT, S. E.; KARAM, A.; PARENT, L. E. Compositional modeling of C mineralization of organic materials in soils. In: INTERNATIONAL SYMPOSIUM OF AGRICULTURAL WASTE MANAGEMENT, 2., 2011, Foz do Iguaçu. **Proceedings...** , Foz do Iguaçu, Brazil, 2011.

PARO, M.; VITTI, G. C.; DONADIO, L. C.; SEMPIONATO, O. R. Influence of
use of two agricultural correctives, limestone and gypsum in the quality of the pear orange fruit. In: CONGRESSO BRASILEIRO DE FRUTICULTURA, 13, 1994,
Salvador. **Annals...** Salvador: Brazilian Society of Fruticulture, 1994. p. 511-512.

PEARSON, K. Mathematical contributions to the theory of evolution - On a form of spurious correlation which may arise when indices are used in the measurement of organs. **Proceedings of the Royal Society of London**, London, v. LX, p. 489-502, 1897.

PEREIRA, F. M.; MARTINEZ JUNIOR, M. **Goiabas para industrialização.**
Jaboticabal: Legis Summa, 1986, 142 p.

PEREIRA, F. M.; NACHITIGAL, J. C. Propagação da goiabeira. In: SIMPÓSIO BRASILEIRO SOBRE A CULTURA DA GOIABEIRA, 1., 1997, Jaboticabal. **Annals...** ,
Jaboticabal: FCAV/Unesp, 1997. p. 17-32.

PRADO, R. M. **Efeitos da calagem no desenvolvimento, no estado nutricional e na produção de fruta da goiabeira e da caramboleira.** 2003. 68 f. Tese (Doutorado em Agronomia) - Faculdade de Ciências Agrárias e Veterinárias, Universidade Estadual Paulista, Jaboticabal, 2003.

PRADO, R. M.; NATALE, W. Calagem na nutrição de calcio e no desenvolvimento do sistema radicular da goiabeira. **Pesquisa Agropecuária Brasileira**, Brasília, v. 39, p. 1007-1012, 2004.

PRADO, R. M.; NATALE, W. Effect of liming on the mineral nutrition and yield of growing guava trees in a typic Hapludox soil. **Communications in Soil Science and Plant Analysis**, Philadelphia, v. 39, p. 2191-2204, 2008.

PRADO, R. M.; NATALE, W.; SILVA, J. A. A. Liming and quality of guava fruit cultivated in Brasil. **Scientia Horticulturae**, Amsterdam, v. 104, p. 1-102, 2005.

QUEIROZ, E. F.; KLIEMANN, H. J.; VIEIRA, A.; RODRIGUES, A. P. M.; GUILHERME,
M. R. Nutrição mineral e adubação da goiabeira (*Psidium guajava* L.). In: HAAG, H. P. (Ed.). **Nutrição mineral e adubação de fruteiras tropicais no Brasil.** Campinas: Cargill Foundation, 1986.

p.165-187.

RAIJ, B. van. **Fertilidade do solo e adubação.** Piracicaba: Ceres, Potafos, 1991. 343 p.

RODRIGUEZ, J. S.; CIBES, N. R.; GONZALEZ IBAÑEZ, J. Deficiency symptons
displayed in guajava (*Psidium guajava* L.) under greenhouse conditions. **University of Puerto Rico: Technical paper**, n. 44, 1968.

ROVIRA, L. A. El ciclo de vida productivo de losfrutales de tipo arbóreo em médio tropical y seus consecuenciasagro-economicas. **Fruits**, Paris, v. 43, p. 517-529, 1988.

ROZANE, D. E.; NATALE, W.; PRADO, R. M.; BARBOSA, J. C. Sample size
Foliar para avaliação do estado nutricional de goiabeiras com e sem irrigação. **Revista Brasileira de Engenharia Agrícola e Ambiental**, Campina Grande, v. 13, p. 233- 239, 2009.

SALVADOR, J. O.; MOREIRA, A.; MURAOKA, T. Deficiência nutricional em mudas de goiabeira decorrente da omissão simultânea de dois macronutrientes. **Pesquisa Agropecuária Brasileira**, Brasília, v. 33, p. 1623-1631, 1998.

SALVADOR, J. O.; MOREIRA, A.; MURAOKA, T. Efeito da omissão combinada de N, P, K e S nos teores foliares de macronutrientes em mudas de goiabeira. **Scientia Agricola**, Piracicaba, v. 56, p. 501-507, 1999.

AGRICULTURE STATISTICAL INFORMATION SERIES: **IEA Yearbook 2000**, São Paulo, v. 12, p. 19, 2001.

SERRANO, L. A. L.; MARINHO, C. S.; RONCHI, C. P.; LIMA, I. M.; MARTINS, M. V.
V.; TARDIN, F. D. Goiabeira 'Paluma' sob diferentes sistemas de cultivo, épocas e intensidades de pruna de frutificação. **Pesquisa**

Agropecuária Brasileira**, Brasília, v. 42, p. 785-792, 2007.

SMITH, G. S.; ASHER, G. J.; CLARK, C. J. **Kiwifruit nutrition:** diagnosis of nutritional disorders. Southern Horticulture, 1985. 56 p.

SOUZA, H. A.; NATALE, W.; PRADO, R. M.; ROZANE, D. E.; ROMUALDO, L. M.; HERNANDES, A. Efeito da calagem sobre o crescimento de goiabeiras. **Revista Ceres**, Viçosa, MG, v. 56, p. 336-341, 2009.

TANNER, J. Fallacy of per-weight and per-surface area standards, and their relation to spurious correlation. **Journal of Applied Physiology**, Bethesda, v. 2, p. 1-15, 1949.

ALL FRUIT. **The culture of the guava.** Available at: < http://www.todafruta.com.br/>. Accessed on: 15 mai. 2012.

WALWORTH, J. L.; SUMNER, M. E. Foliar diagnosis: a review. **Advances in Plant Nutrition**, New York, v. 539, p. 193-240, 1988.

WELTJE, G. J. Quantitative analysis of dentrial modes: statistically rigorous confidence regions in ternary diagrams and their use in sedimentary petrology. **Earth Science Reviews**, [S.I.], v. 57, p. 211-253, 2002.

CHAPTER 2 - Soil cation balance in guava orchards

To reach high yields, *guava* (*Psidium guajava* L.) requires quantities of nitrogen and potassium fertilizers that are compatible with the estimated use of these nutrients, taking into account the availability of these elements in the soil. However, the concept of base saturation of the cation exchange capacity (CEC), used for the evaluation of soil fertility, is biased because soil data can transmit relative information and interact within the restricted space of the CEC, generating spurious correlations. Thus, the objective of the work was to evaluate a system of isometric log ratios (*ilr*) to correct such distortions, enabling the management of nitrogen and potassium fertilization in guava orchards. Data from experiments conducted during three years were used, being two distinct experiments with N and K doses, in Argissolo and Latosolo. The values of K, Ca, Mg and of potential acidity (H + Al) were divided into three binary partitions as follows: the soil balance [K | Ca, Mg, H + Al] to manage K fertilization; the soil balance [Ca, Mg | H + Al] to monitor the need for lime; and, the soil balance [Ca | Mg] to establish liming material. Nitrogen fertilization decreased the soil [Ca, Mg | H + Al] balance due to soil acidification. Potassic fertilization increased the soil balance [K
| Ca, Mg, H + Al]. The critical values of the soil balances were -2.10 for [K | Ca, Mg, H + Al], 1.28 for [Ca, Mg | H + Al] and 0.85 for [Ca | Mg], since high yields in guava production (> 77 t ha-1) were obtained with balance values above those de The percentages of CTC saturation were 2.0% K, 25% Ca, 7.5% Mg and 65% (H + Al). The critical pH value (CaCl2) for the guava crop was 4.5. Maintenance of the soil balance [K | Ca, Mg, H + Al] requires that K exports from the crop be compensated by equivalent amounts of potassium fertilizers. The critical soil cation balances were defined without numerical trends and are therefore an optimal soil fertility index for fertilization management in guava orchards.

Keywords: compositional data analysis, soil cation balance, soil fertility, Aitchison geometry, *Psidium guajava* L., isometric log ratios (*ilr*)

Introduction

Guava (*Psidium guajava* L.) requires substantial amounts of N and K to achieve high productivity and fruit quality (ARORA; SINGH, 1970; ANJANEYULU; RAGHUPATHI, 2009; OSMAN; ABD EL-RAHMAN, 2009). However,
In Brazil, fertilizer recommendations for guava are based on a limited number of field trials and soil and leaf analysis.

The interpretation of soil analyses is based on the formation and maintenance of sufficient levels of available nutrients (NSND) (McLEAN, 1984). The balance of nutrients is not part of the soil analysis procedures, despite the infinity of existing interactions between the elements that occur in the plant during and after nutrient uptake (MALAVOLTA, 2006).

The concept of nutrient intensity and balance (CIEN) is a system for interpreting the ionic balance between soluble elements and total soil salt concentrations (GERALDSON, 1957; 1970; 1984). On the other hand, the concept of cationic base saturation rate (TSBC), uses dual relations of exchangeable cations, which exchange at cation exchange capacity (CEC), such as K/Ca, K/Mg and Ca/Mg (McLEAN et al., 1983; McLEAN, 1984). The latter concept has been criticized due to the scarcity of experimental evidence on what the "ideal" soil cationic relationships would be (LIEBHARDT, 1981; KOPITTKE; MENZIES, 2007). The NSND, CIEN and TSBC approaches are conceptually biased as they are intrinsically influenced by scale dependence, redundancy and non-normal distribution of the data since their database is compositional (i.e. they are expressed as percentages).

In fact, soil cation data are strictly positive multivariate data, being part of the closed compositional space of CTC (PARENT, 2011). Similar to the soil texture diagram, K-Ca-Mg relationships can be illustrated by ternary diagrams, where one component is calculated by the difference between 100% and the sum of the other two. Thus, the third component will always be redundant; so, to avoid this redundancy, the diagnosis should be performed using D-1 degrees of freedom (EGOZCUE; PAWLOWSKY-GLAHN, 2005). The cationic relations are not a solution, since among the K/Ca, K/Mg and Ca/Mg relations, derived from the

compositional vector K-Ca-Mg, the third component can be determined from the other two, e.g. K/Ca = [(K/Mg) / (Ca/Mg)].

The problem of spurious correlations between component proportions and proportions relative to a measurement scale (such as dry matter, fresh matter or organic basis), is well known and was previously observed by Pearson (1897), Tanner (1949) and Chayes (1960). For example, the results of a multivariate analysis will vary either way, no matter whether the concentration of the cationic species is expressed on a molar or mass basis. Since they affect the covariance matrix, spurious correlations will distort linear statistical models and may even produce contradictory results (PARENT et al., 2011).

Compositional data exhibit non-normal distribution, within its closed space between 0 and 100%, since the lower and upper limits of the confidence interval cannot be negative or greater than 100% (DIAZ-ZORITA et al., 2002; WELTJE, 2002). However, log ratios can assume negative or positive values, varying freely in real space ± ∞. Aitchison (1986) suggested using additive log ratio (*alr*) and centralised log ratio (*clr*) transformations for compositional data. The centralized log ratio has been used in studies on plant nutrition (PARENT; DAFIR, 1992; PARENT et al., 2009), soil solution (PARENT et al., 1997) and hydroponics (LOPEZ et al., 2002). Egozcue et al. (2003) developed isometric log ratio transformations (*ilr*), based on D-1 balances, arranged orthogonally. This approach is unbiased, unbiased, (D-1 degrees of freedom), and preserves all the information contained in the compositional vector thanks to the orthogonality principle. The *ilr* was found to be the most suitable method for conducting multivariate analyses, as demonstrated in Filzmoser and Hron (2011). The *ilr* concept has been successfully tested in geosciences (BUCCIANTI, 2011), as well as in studies of plant nutrition (PARENT, 2011), organic waste decomposition (PARENT et al., 2011) and soil aggregation (PARENT et al., 2012).

However, an unbiased, mathematical concept for diagnosing soil cation balance has never been developed before.

Thus, the hypothesis of the present study is that the diagnosis of soil cation balance using the *ilr* concept is better than the diagnosis resulting from the use of other approaches. With this, the objective of this work was to monitor the changes in soil cation balance, arising from nitrogen and potassium fertilization, in guava orchards of the cultivars Rica and Paluma, using the unbiased *ilr* concept. A scheme of cation balances arranged to facilitate the management of K fertilizers and liming materials containing Ca and Mg in guava orchards was tested using soil analysis data and estimated nutrient utilization.

Material and methods

Data from experiments conducted with the guava crop by Natale (1993) and Natale (1999), collected during the growth and development of the crop were used as a database for the current study, with 260 observations.

Experimental area and treatments

The experiments were conducted in two different guava orchards, during three consecutive agricultural seasons, from 1989 to 1992. The fertilization experiments were installed in Jaboticabal (São Paulo), on a Red-Yellow Argissolo Vermelho-Amarelo, clay of low activity, eutrophic, abrupt, A moderate, sandy/clayey texture (EMBRAPA, 2006) and, in São Carlos (São Paulo), on a Red-Yellow Latosol, epieutrophic, endodistrophic, A moderate, medium texture, cerrado phase and flat relief (EMBRAPA, 2006). The orchards were formed by one-year-old guava trees 'Rica' in Jaboticabal and 'Paluma' in São Carlos, two important guava cultivars in Brazil (PEREIRA, 1984). Both cultivars are early, vigorous and productive. The fertilizer doses of the treatments were distributed in the first year of plant establishment and increased proportionally each year, taking into account the removal of nutrients by the harvest and, principally, the increase in the demand for nutrients due to the development of the fruit trees.

Separate experiments were conducted in both areas, one with nitrogen fertilization and the other with potassium fertilization. The area of each experiment was 3,360 m^2 in Jaboticabal and 3,920 m^2 in São Carlos. The experimental design was in randomized blocks with six treatments ('Rica') and seven treatments ('Paluma'), all with four repetitions. The plots consisted of four plants, with spacing of 7 m by 5 m, and the first plant represented the border, considering only three plants as useful for evaluation purposes. The conduction system was of cleaning pruning and conduction, with one production per year, non-irrigated.

In the N experiment, the treatments in the first year were zero, 30, 60, 120, 180 and 240 g of N plant-1 in Jaboticabal and zero, 30, 60, 120, 180, 240 and 300 g of N plant-1 in São Carlos. The nitrogen fertilization was supplemented with 120 g of P2O5 plant-1 and 120 g of K2O plant-1. The initial doses of N were doubled and tripled in the second and third years, respectively. The initial doses of P and K were also doubled in the second year. In the third year, the doses were 240 g of P2O5 plant-1 and, 360 g of K2O plant-1. The sources used were ammonium nitrate (34% N), simple superphosphate (18% P2O5) and potassium chloride (60% K2O).

In the K experiment, the treatments in the first year were zero, 30, 60, 120, 180 and 240 g of K2O plant-1 in Jaboticabal and zero, 30, 60, 120, 180, 240 and 300 g of K2O plant-1 in São Carlos, together with 120 g of N plant-1 and 120 g of P2O5 plant-1. The doses of N, P and K were doubled in the second year. The doses of K were tripled in the third year and supplemented with 360 g of N plant-1 and 240 g of P2O5 plant-1. The sources used were ammonium sulphate (20% N), triple superphosphate (44% P2O5) and potassium chloride (60% K2O).

Fertilizers were applied manually in a 40 cm strip in the crown projection. While the experiments were being conducted in the field, the normal cultural treatments necessary for the good development of the plants were carried out. To evaluate the production, the fruits were harvested one to three times a week, from January/February to May/June of each year.

Sampling and soil chemical analysis

The soils were sampled annually after harvest, in four sub-samples per plant (North, South, East and West), in the 0-20 cm layers (during the three years of the experiment) and 20-40 cm layers (only in the last two years), where most of the guava root system is located (FRACARO; PEREIRA, 2004), for a total of 12 sub-samples per plot, in order to constitute the composite sample. The soil samples were air dried and analyzed for pH (CaCl2), organic matter content, K, Ca, Mg and H + Al (RAIJ et al., 1987). The K, Ca and Mg were extracted by exchange resins and quantified by absorption spectrophotometry

were expressed in mmolc dm-3. The potential acidity (H + Al) was quantified by the pH SMP buffer method (SHOEMAKER et al., 1961) and, by the equation of Quaggio et al. (1985), converted into mmolc (H + Al) dm-3:

$$(H + Al) = 10 \exp (7.76 + 1.053 \, pHSM); \; R2 = 0.98 \qquad (1)$$

The cation exchange capacity (CEC) was calculated by summing the cationic species. The CEC was 54 mmolc dm-3 in the Latosol ('Paluma') and 41 mmolc dm-3 in the Argissolo ('Rica'), while the organic matter content was 19 and 22 g dm-3, respectively, at the beginning of the experiments.

Soil cation balance

The compositional space (S) of the exchangeable species was defined as follows
form:

$$SD = C \, (K, Ca, Mg, H + Al) \qquad (2)$$

where D = four components and C is the function closure operator with reference to CTC.

Relationships between cationic species were projected into Euclidean space using D-1 ilr coordinates with orthonormal basis, calculated as follows (EGOZCUE et al., 2003):

$$ilr_j = \sqrt{\frac{rs}{r+sg(c-)}} \, \ln \frac{g(c+)}{} \; \text{with } j = [1,2, \dots , D - 1] \qquad (3)$$

where: ilr_j is the isometric j-th log ratio, r and s represent the numbers of components in positive (+) and negative (-) groups in each binary partition, respectively, $g(c+)$ is the geometric mean of the components in the positive groups

$c+$, and $g(c-)$ is the geometric mean of the components in the negative groups $c-$ (Table

1). The coefficient $\sqrt{\frac{rs}{r+s}}$ represents the number of components in the positive groups and

negative. As κ is the unit of measurement, the relationships of the parts of the groups in equation 2 eliminates the κ, making the scale of the *ilr* coordinates invariant and free of spurious correlations caused by scale dependence. Positive values of *ilr* indicate that the positive group classified as positive has greater weight than the other group, while null values indicate the same weight.

Table 1. Sequential binary partitioning of the CEC components in mmolc

Ilr	K	Ca	M g	H+ Al	Balance	r†	S	*ilr* calculation
					Sequential binary partition			
1	1	-1	-1	-1	K | Ca, Mg, H + Al]	1	3	$\sqrt{\frac{1x3}{1+3}}\, ln\left(\frac{K}{[CaxMgx(H+Al)]^{1/3}}\right)$
2	0	1	1	-1	Ca, Mg | H + Al]	2	1	$\sqrt{\frac{2x1}{}}^{1/2} ln\left(\frac{(CaxMg)}{}\right)$
3	0	1	-1	0	Ca | Mg]	1	1	$\sqrt{\frac{1x1}{}}\, ln\left(\frac{Ca}{}\right)$

†r = number of positive signs and *s* = number of negative signs.

The binary sequential partition (PBS) in Table 1 was prepared to facilitate interpretation of the results. Potassium fertilization was balanced against Ca, Mg and potential acidity. An alternative soil PBS could be a contrast between cationic species and potential acidity, i.e. [K, Ca, Mg | H + Al], followed by the soil balances [K | Ca, Mg] and [Ca | Mg]. This alternative PBS is a balance representation of the classical K/Mg, K/Ca and Ca/Mg ratios, where the potassium ratios are compressed into the soil balance [K | Ca, Mg]. The choice of PBS depends on the objective of the study. The PBS chosen in Table 1 was selected to highlight the potassium balance and, the Ca and Mg requirements, to determine the need for liming, due to acidification by nitrogen fertilization, mainly.

As an example, a soil presenting 65% Ca, 10% Mg, 5% K and 20% total acidity (McLEAN, 1984) has three *ilr* coordinates (four components minus one), calculated as follows:

$$ilr1 = \sqrt{\frac{r}{r+} } \; ln \frac{rsK1x35}{s[CaxMgx(H+Al)]^{1/3}} = \sqrt{ln} \frac{}{1+3} = -1{,}341 \frac{}{(65x10x20)^{1/3}} \tag{4}$$

where $r = 1$ and $s = 3$.

The second coordinate contrasts divalent cations and total acidity, calculated as follows:

$$ilr2 = \sqrt{\frac{r}{r+}} \; ln \frac{rs(CaxMg)^{1/2}}{s(H+Al)} \frac{2x1}{2+120} = \sqrt{ln} \frac{(65x10)^{1/2}}{} = 0{,}198 \tag{5}$$

The third coordinate [Ca | Mg] maintains orthogonality and is calculated as follows:

$$ilr3 = \sqrt{} \; ln \frac{rsCa1x165}{r+sMg1+110} = \sqrt{ln} = 1{,}324 \tag{6}$$

Calculations

The use of nutrients between the sampling periods was estimated as the amount of nutrients added in fertilization, less the amount of nutrients exported by the harvest. According to Natale et al. (2002), each tonne of fresh fruit from the Rica cultivar removes 1,325 g N, 166 g P and 2,180 g K, while the Paluma cultivar removes 1,179 g N, 121 g P and 1,554 g K.

Statistical analyses

The *ilr* calculations, discriminant analyses and spurious correlation analyses were conducted in the R statistical computing environment (IHAKA; GENTLEMAN,

1996), which is a free *software,* using the compositional package (van den BOOGAART; TOLOSANA-DELGADO, 2008) for compositional data analysis. Spurious correlations were identified by comparing soil volume (mmolc dm- ^{3}) or TCC (% saturation) as measurement scales for cationic species.

Analyses of variance were conducted using PROC MIXED of SAS (*Statistical Analysis System*, version 9.2., 2009), and their significance was checked by the t-test.

Critical *ilr* values were obtained using the Cate- Nelson procedure, by dividing the data points of the graph relating fruit production and their corresponding *ilr* values (NELSON; ANDERSON, 1984). Briefly, the data points were distributed in four quadrants, using crossed lines perpendicular and parallel to the axes, which were moved across the graph until the number of points in opposite quadrants was maximized or minimized. Samples from the points in the quadrants were interpreted with reference to nutrient imbalance, as follows (GLYNN; WEISBACH, 2012):

- True positive (VP: nutrient imbalance): unbalanced crop (low yield), correctly diagnosed as unbalanced (above the critical value);
- False positive (FP: type I error): balanced crop (high yield), incorrectly identified as unbalanced (above the critical value);
- True negative (VN: nutrient balance): balanced crop (high yield), correctly diagnosed as balanced (below the critical value);
- False negative (FN: type II error): unbalanced crop (low yield), incorrectly identified as balanced (below the critical value).

The critical soil balances were re-transformed to percentages using the R-compositional package and, for the initial concentration units (mmolc dm-3), using the CEC values. As there were three soil balances and four components, the calculation procedure automatically sets the sum of the cation species at 100% CTC saturation to solve a system of 4 equations with 4 unknowns (the 3 *ilrs* and the restriction sum).

Results and Discussion

Study of correlations

The cationic species, expressed on a soil volume basis (SVB) or as saturation CTC, showed evidence of spurious correlations (Tables 2 and 3).

Table 2. Spurious correlations between soil cations expressed in mmolc dm-3

	K	Ca	Mg
K	-	0,54**	- 0.01ns
Ca	-	-	0,44*
Mg	-	-	-

* significant at 0.05 probability; ** significant at 0.01 probability, ns: not significant.

48

Table 3. Spurious correlations between soil cations expressed on the basis% of saturation of CTC

	%K	%Ca	%Mg
%K	-	0,48*	-0,18ns
%Ca	-	-	0,22ns
%Mg	-	-	-

* significant at 0.05 probability; ** significant at 0.01 probability, ns: not significant.

The correlation coefficient between K and Ca was 0,54** in the BVS base and 0,48* in the CTC base. The correlation coefficient between K and Mg was -0.01ns on the BVS basis and -0.18ns on the CTC basis. Finally, the correlation coefficient between Ca and Mg was 0.44* on the BVS basis and 0.22ns on the CTC basis. Thus, the correlation coefficients varied in magnitude and significance depending on the scale of measurement, indicating spurious correlations that may lead to different interpretations.

An additional advantage of the *ilr* concept is scale invariance. Mehlich (1973) raised the problem of volumetric or gravimetric measurement scales in soil analysis and, the methodological bias introduced by converting scales using a single bulk density value. The *ilr* technique is therefore an effective means of correcting the scale dependence problem.

Fruit production and soil cation balance

Fruit production was significantly ($P \leq 0.05$) influenced by year, doses and their interactions, in both areas, while soil nutrient balances were significantly ($P \leq 0.05$) influenced by year and fertilizer doses (Tables 4 to 7).

In the experiment with N doses, production increased linearly in the second (y = 0.0656x + 17.824; R^2 = 0.90) and third years (y = 0.1456x + 27.189; R^2 = 0.90) in the Rica cultivar (Figure 1a) and, only in the third year (y = 0.0486x + 63.810; R^2 = 0.85), in the Paluma cultivar (Figure 1b). In the experiment with doses of K, the production increased in a quadratic manner in the third year with the Rica cultivar (y = 0.0003x2 + 0.0201x + 36.509; R^2 = 0.94) (Figure 2a) and linearly in the third year (y = 0.0716x + 65.260; R^2 = 0.73), with the Paluma cultivar (Figure 2b). The production increased annually until it reached its maximum value in the third year after crop establishment. Thus, it is justified that the doses of N and K were adjusted annually during the experiment, in order to consider exports through the harvest and, principally, to supply the demand of the plant due to its development.

The Paluma cultivar was more productive than 'Rica', and in Brazil many 'Paluma' orchards produce more than 100 t guava ha-1 and are therefore expanding.

Table 4. Analysis of variance and F values on the effect of N addition on yield and soil cation balance in a 'Rica' guava orchard

Year Production		K | Ca, Mg, H+Al] Mg][Ca |	[Ca,Mg | H+Al][Mg]	Ca | Mg]['Rich'	K | Ca, Mg, H+Al][Ca		,Mg | H+Al][Ca |
Year (A)	-249**5	,69ns	0,01ns	9,16ns	6,58ns	5,01ns	
(D)	4,39ns Doses -33,	4**9,	81**20,	48**12,	3**18,	04**7,	36**8,60**
A x D	-7 ,61* *0	,67ns	1,20ns	1,76ns	1,61ns	0,31ns	0.53ns
	1ns						
	28,08**nsnsnsnsnsns						
	339,7**						
C vs. T†	1ns						
	223,9*	*25,	22**15,	57**7,	99**61,	48**nsns	
	3116**						
Setting	1ns						
	2L††	QQLQQL					
	3L						
C.V. (%)	16,7	-1,	2-10,	01,	8-0,	9-6,	35,5

† C vs. T = Control (zero dose) vs. fertilized treatments. †† Adjustment L = linear, Q = quadratic.
* significant at 0.05 probability; ** significant at 0.01 probability, ns: not significant.

Table 5. Analysis of variance and F values on the effect of N addition on yield and soil cation balance in a Paluma guava orchard

	Year Production	Layer 0-20 cm			Layer 20-40 cm			
		K	Ca, Mg, H+Al] Mg]	[Ca, Mg	H+Al][	Ca	Mg][	K	Ca, Mg, H+Al]	[Ca, Mg	H+Al][Ca	Mg]	[Ca	
		'Paluma'						
Year (A)	-	641**	1,26ns	146,76**	26,7**	0.46ns	20,39*	10,77*
Dose (D)	-	2.16ns	0.92ns	20,36**	1.91ns	2,42*	3,54**	3,35*
A x D	-	2,36**	0.13ns	2.25ns	2.03ns	0,28ns	0.54ns	2.02ns
	1	ns						
	2	ns	ns	ns	ns	ns	ns	ns
	3	5,57**						
C *vs*. T†	1	ns						
	2	ns	ns	24,68**	ns	ns	ns	ns
	3	7,15**						
Setting	1	ns						
	2	ns	ns	L††	ns	ns	L	L
	3	L††						
C.V. (%)		14,6	-1,1	-1,9	22,8	-1,4	-2,3	45,0

† C *vs*. T = Control (zero dose) *vs.* fertilized treatments. †† Adjustment L = linear, Q = quadratic.
* significant at 0.05 probability; ** significant at 0.01 probability, ns: not significant.

Table 6. Analysis of variance and F values on the effect of K addition on yield and soil cation balance in a 'Rica' guava orchard

Year	Production	Layer 0-20 cm			Layer 20-40 cm		
		K | Ca, Mg, H+Al]	Ca, Mg| H+Al]	Ca | Mg]	K | Ca, Mg, H+Al]	Ca, Mg| H+Al]	Ca | Mg]
				'Rich'			
Year (A) -	195**	4.75ns	4.00ns	4,25ns	1.37ns	1.76ns	1.86ns
Dose (D) -	4,32**	25,38**	1.21ns	0.91ns	20,45**	1.09ns	1,22ns
A x D -	3,58**	1.64ns	0.68ns	1.11ns	0.63ns	0.68ns	0.58ns
1	ns						
2	ns	ns	ns	ns	ns	ns	ns
3	10,9**						
C vs. T† 1	ns						
2	ns	53,06**	ns	ns	32,45**	ns	ns
3	16,5**						
Setting 1	ns						
2	ns	L	ns	ns	L	ns	ns
3	Q††						
C.V. (%)	17,8	-1,9	-3,2	1,6	-1,4	-5,4	4,0

† C *vs.* T = Control (zero dose) *vs.* fertilized treatments. †† Adjustment L = linear, Q = quadratic.
* significant at 0.05 probability; ** significant at 0.01 probability, ns: not significant.

Table 7. Analysis of variance and F values on the effect of adding K on yield and soil cation balance in a Paluma guava orchard

	Year	Production	Layer 0-20 cm			Layer 20-40 cm		
			K | Ca, Mg, H+Al]	Ca, Mg| H+Al]	Ca | Mg]	K | Ca, Mg, H+Al]	Ca, Mg| H+Al]	Ca | Mg]
					'Paluma'			
Year (A)	-	258**	4.72ns	60,22**	5,21*	0.10ns	7.57ns	0,26ns
Dose (D)	-	3,66**	47,67**	1,19ns	0,20ns	48,96**	0.76ns	0.76ns
A x D	-	5,25**	1.46ns	0.79ns	0.82ns	2.29ns	0.60ns	0.37ns
	1	ns						
	2	ns	ns	ns	ns	ns	ns	ns
	3	13,14**						
C *vs.* T†	1	ns						
	2	ns	131,38**	ns	ns	78,61**	ns	ns
	3	53,62**						
Setting	1	ns						
	2	ns	L	ns	ns	L	ns	ns
	3	L††						
C.V. (%)		14,1	-1,0	-2,4	5,1	-1,0	-1,4	10,3

† C *vs.* T = Control (zero dose) *vs.* fertilized treatments. †† Adjustment L = linear, Q = quadratic.
* significant at 0.05 probability; ** significant at 0.01 probability, ns: not significant.

response to the addition of N. DMS = minimum significant difference at 0.05 probability. [†] N doses in the 1° year: zero, 30, 60, 120, 180 and 240g of N plant-1, in the 2° and 3° years double and triple the initial doses respectively.

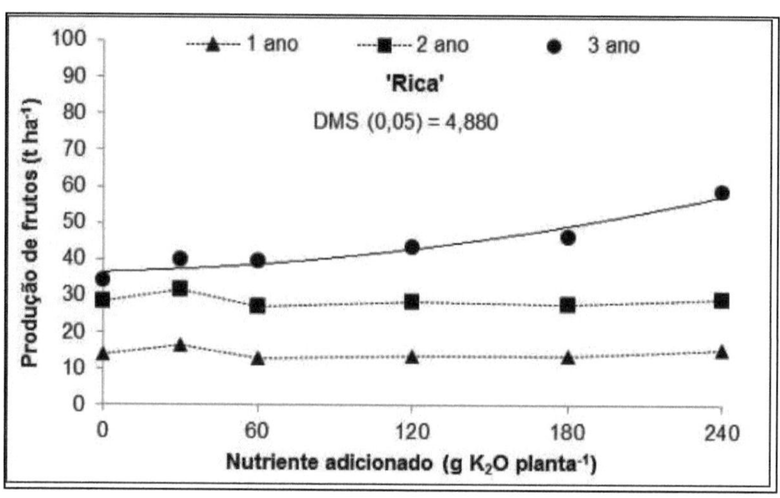

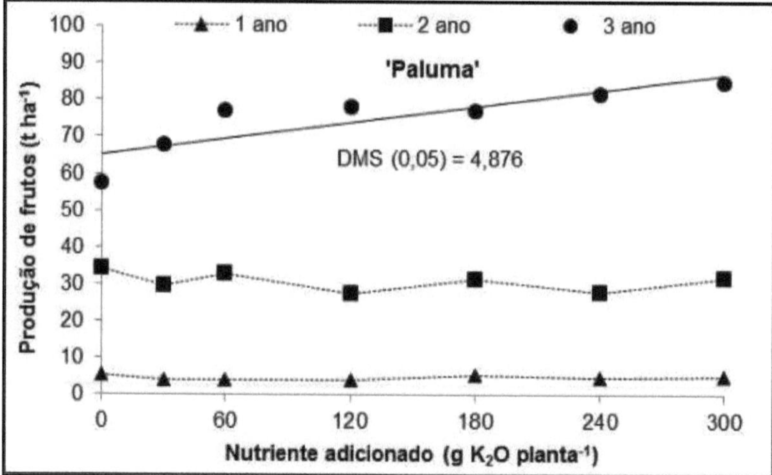

Production of guava from Rica (2a) and Paluma (2b) cultivars, in response to the addition of K. DMS = minimum significant difference at 0.05 probability. [†] Doses of K in the 1° year: zero, 30, 60, 120, 180, 240 and 300g of K2O plant-1, in the 2° and 3° years double and triple the initial doses, respectively.

The treatments with N doses decreased in a quadratic way the soil balance [K | Ca, Mg, H + Al] in the 0-20 cm layer ($y = 2*10-5x2 - 0.0061x - 1.1582$; $R^2 = 0.81$) and 20-40 cm ($y = 2*10-5x2 - 0.0062x - 1.3428$; $R^2 = 0.82$) (Figure 3a), and the soil balance [Ca, Mg | H + Al] in the 0-20 cm layer ($y = 2*10-5x2 + 0.0011x - 0.3332$; R^2

= 0.91) and 20-40 cm (y = $2*10^{-5}x^2$ + 0.0036x - 0.3668; R^2 = 0.61) (Figure 3b), and linearly increased the soil balance [Ca | Mg] in 'Rica' in the 0-20 cm (y = 0.002x + 0.9112; R^2 = 0.86) and 20-40 cm (y = 0.0006x + 0.7688; R^2 = 0.38) layers (Figure 3c). In 'Paluma', the treatments with N doses linearly decreased the soil balance [Ca, Mg | H + Al] in the 0-20 cm layer (y = -0.0022x - 0.7665; R^2 = 0.96) and 20-40 cm (y = -0.001x - 1.029; R^2 = 0.69) (Figure 3b), and linearly increased the soil balance [Ca | Mg] in the 20-40 cm layer (y = 0.0006x + 0.3346; R^2 = 0.48) (Figure 3c). On the other hand, treatments with K doses linearly increased the soil balance [K | Ca, Mg, H + Al] in the 0-20 cm layer (y = 0.0048x - 2.2091; R^2 = 0.98) and 20-40 cm (y = 0.0048x - 2.5119; R^2 = 0.97) in 'Rica' orchards (Figure 4a), and in the 0-20 cm (y = 0.0036x - 2.6028; R^2 = 0.95) and 20-40 cm (y = 0.0047x - 2.7531; R^2 = 0.99) in 'Paluma' orchards (Figure 4a). Nitrogen fertilization decreased soil pH (Figure 5a), since N-NH4 acidifies the rhizosphere, due to nitrification or the absorption of ammonium by the roots, but stimulated plant growth and, consequently, nutrient uptake, leading to a decrease in K, Ca and Mg in the soil (Figures 5b, 5c and 5d). The treatments with doses of potassium did not alter the pH (Figure 6a), but increased the concentration of K in the soil (Figure 6b) and maintained the concentrations of Ca and Mg in the soil (Figures 6c and 6d). Thus, the estimated utilization of K (added K minus exported K) was negative (Figure 7a) or balanced (Figure 7b) in the experiment with N and positive in the experiment with K (Figures 8a and 8b).

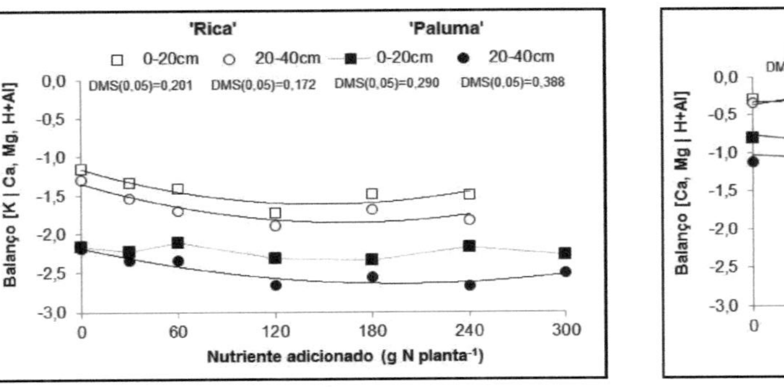

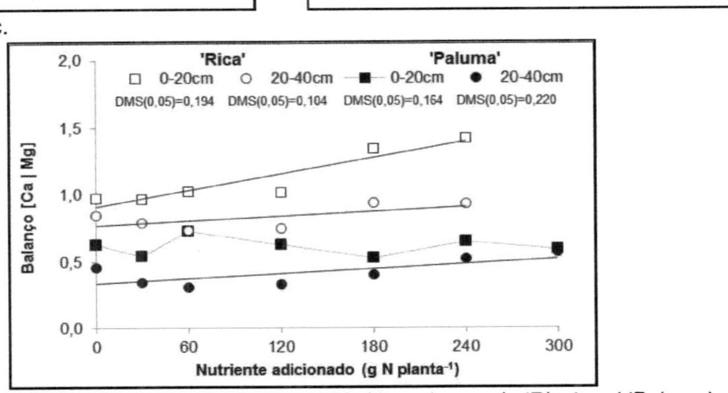

Figure 3. Seasonal change in soil cation balances in the 0-20 and 20-40 cm layers, in 'Rica' and 'Paluma' guava orchards, in experiment with N doses. DMS = minimum significant difference at 0.05 probability. † N doses in the 1° year: zero, 30, 60, 120, 180, 240 and 300 g of N plant-1, in the 2° and 3° years double and triple the initial doses, respectively.

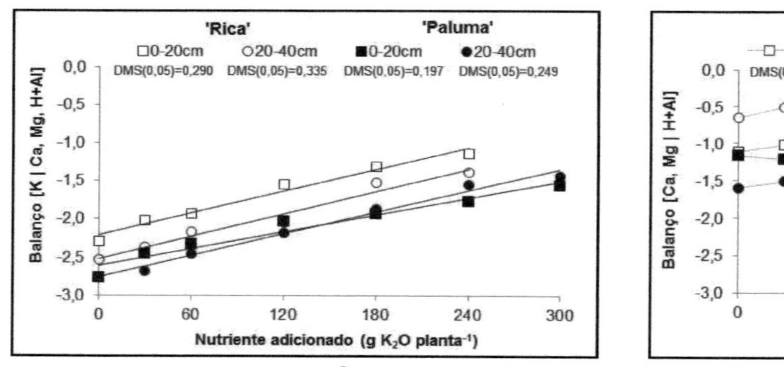
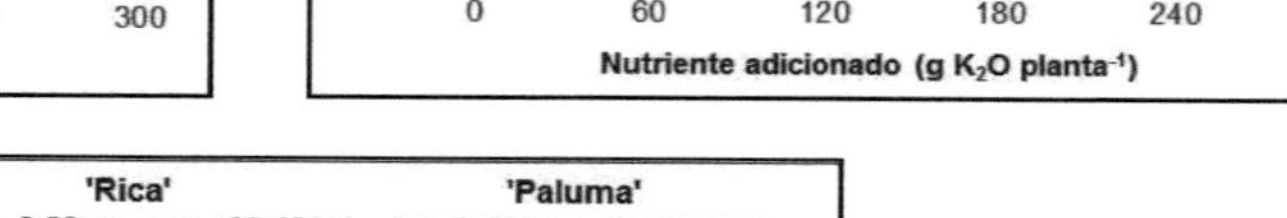
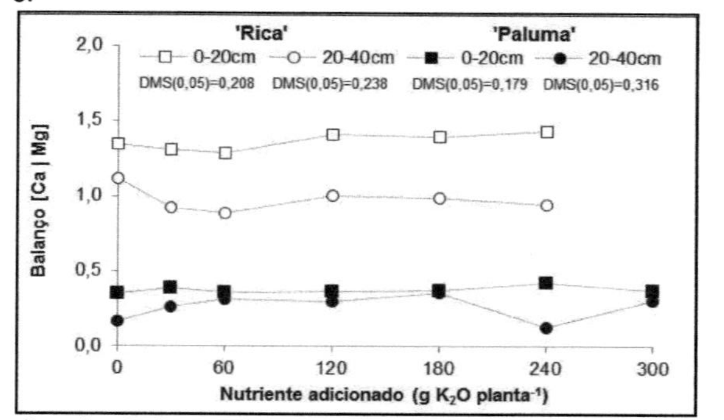

Figure 4. seasonal changes in soil cation balances in the 0-20 and 20-40 cm layers, in 'Rica' and 'Paluma' guava orchards, in an experiment with K doses. DMS = minimum significant difference at 0.05 probability. [†] K doses in the 1° year: zero, 30, 60, 120, 180, 240 and 300 g of K2O plant-1, in the 2° and 3° years double and triple the initial doses, respectively.

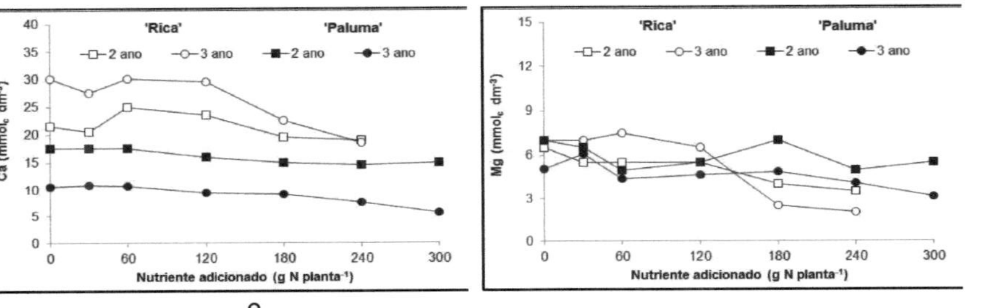
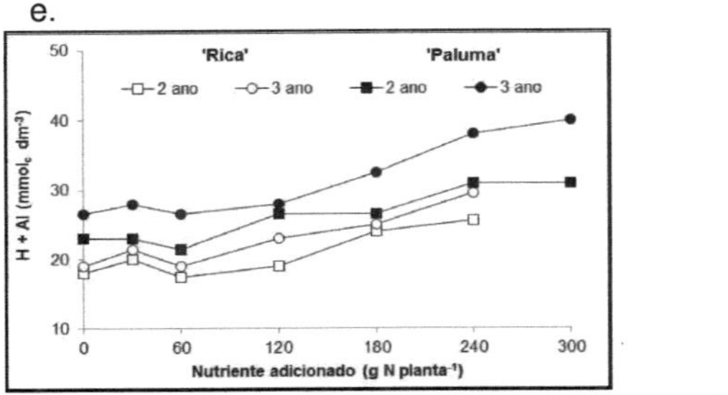

e.

Figure 5. Seasonal change in median values of soil chemical properties, in 'Rica' and 'Paluma' guava orchards, in experiment with N doses. † N doses in the 1° year: zero, 30, 60, 120, 180, 240 and 300 g of N plant-1, in the 2° and 3° years double and triple the initial doses, respectively.

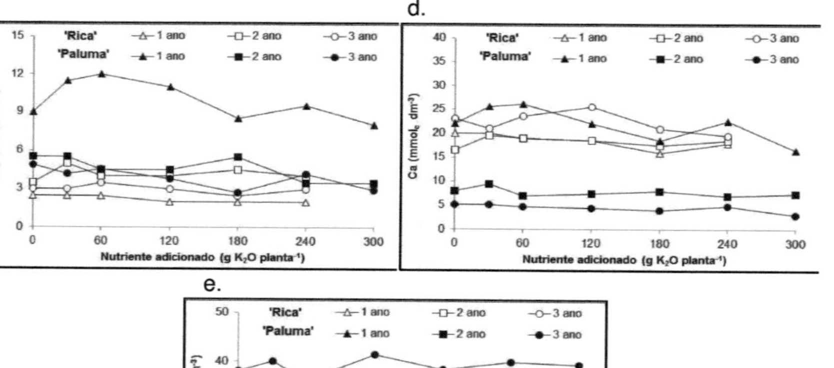
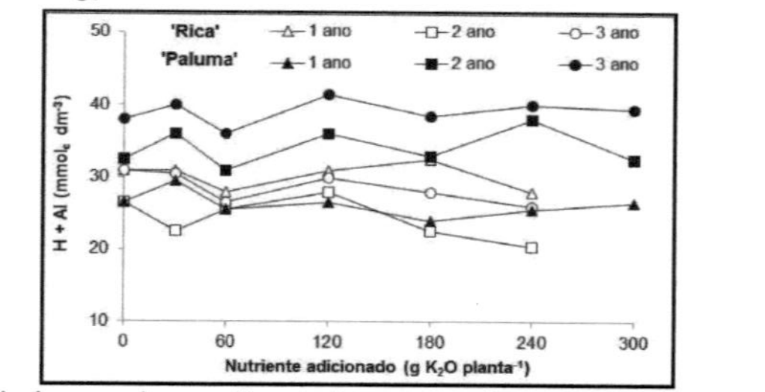

Figure 6. Seasonal change in median values of soil chemical properties in 'Rica' and 'Paluma' guava orchards in an experiment with K doses. [†] K doses in the 1° year: zero, 30, 60, 120, 180, 240 and 300 g K2O plant-1, in the 2° and 3° years double and triple the initial doses, respectively.

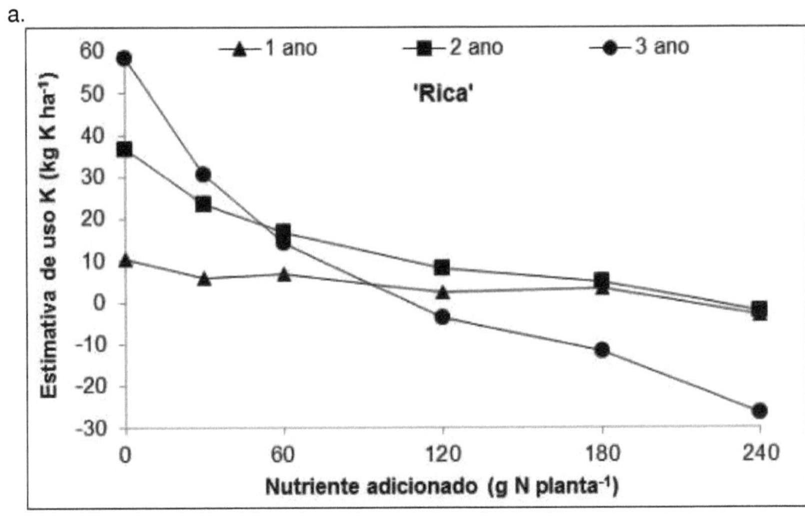

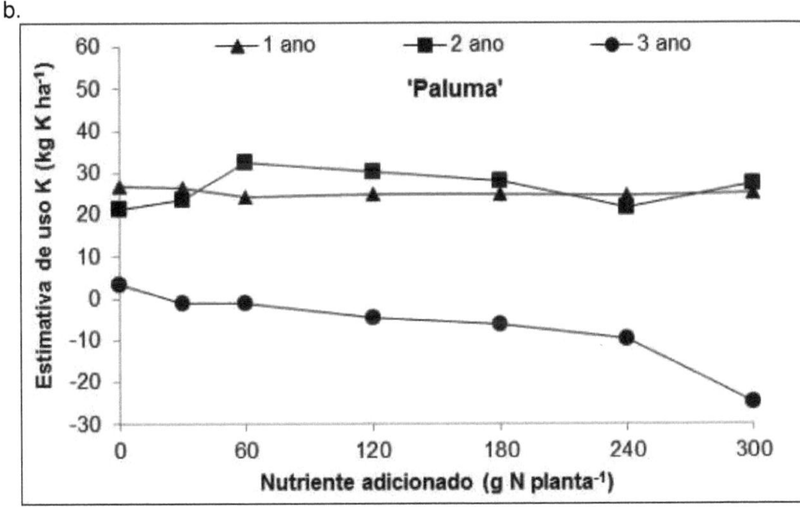

Figure 7. Potassium removal by 'Rica' (7a) and 'Paluma' (7b) guava trees in the experiments employing N doses. † N doses in the1° year: zero, 30, 60, 120, 180, 180, 240 and 300 g of N plant-1 and, in the 2° and 3° years double and triple the initial doses, respectively.

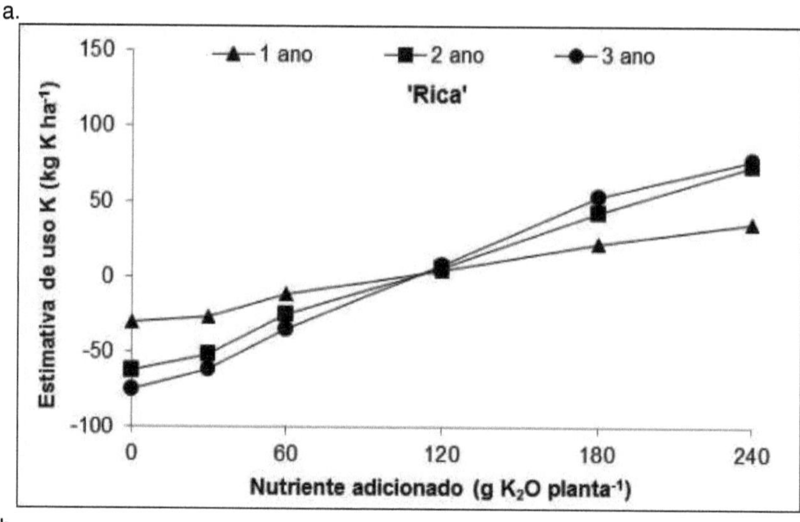

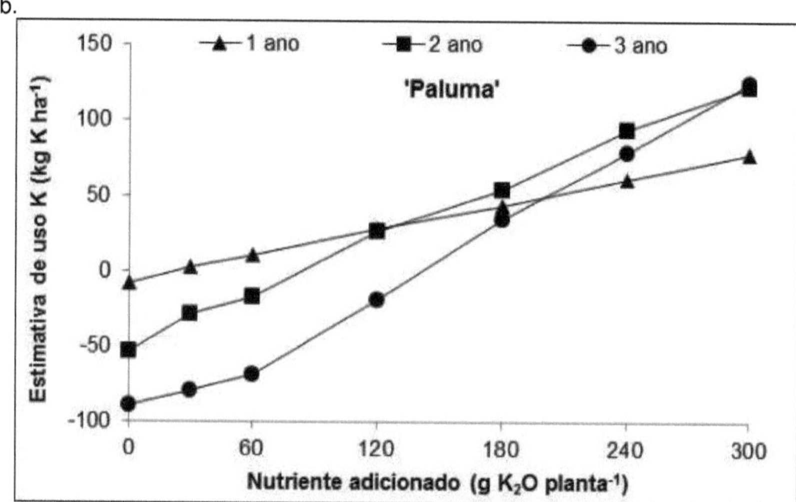

Figure 8. Potassium removal by 'Rica' (8a) and 'Paluma' (8b) guava trees in the experiments employing K doses. † K doses in the 1° year: zero, 30, 60, 120, 180, 240 and 300 g of K2O plant-1, in the 2° and 3° years double and triple the initial doses, respectively.

N stimulates plant growth and consequently the uptake of K, since K is the principal cation that accompanies nitrate, besides being the regulator of several plant growth processes (MALAVOLTA, 2006). On the other hand, cation balances in the soil are related to the magnitude of the mechanisms that restrict the availability of each ion, the quantity of the ion and the bond strength. Ammonia and acidity, resulting from the nitrification process, compete with K, Ca and Mg for adsorption sites on clay particles (MALAVOLTA, 1989). The exchangeable cations are located mainly at the ends and at the base (OH) of the kaolinite surface, where permanent negative charges are insignificant and the CEC depends strongly on particle size and pH (MA; EGGLETON, 1999). The high surface charge density of kaolinite increases the dehydration of K, resulting in a high affinity of kaolinite for K, in addition to a preference for K over Ca (LEVY et al., 1988). Thus, there is a limited range of acid-base equilibrium in kaolinite that can make K, Ca and Mg available.

Dendograms and ternary diagrams of compositional values

The soil cation balances in the 0-20 and 20-40 cm layers can be compared using dendograms, ternary diagrams and Euclidean representation of the relationship between the soil balances [K | Ca, Mg, H + Al] and [Ca, Mg | H + Al] which, by definition, are orthogonal to each other.

In the dendograms of soil balances, illustrated in Figure 9, each intersection is a weighted balance, defined from the sequential binary partition shown in Table 1. The balance can be zero (centre), negative (shifted to the left) or positive (shifted to the right). The percentage variance of each balance is represented by the length of the vertical bar. The soil cation balances were similar in the 0-20 and 20-40 cm soil layers. The soil potassium balance was the most prominent being driven by the K-dose treatments.

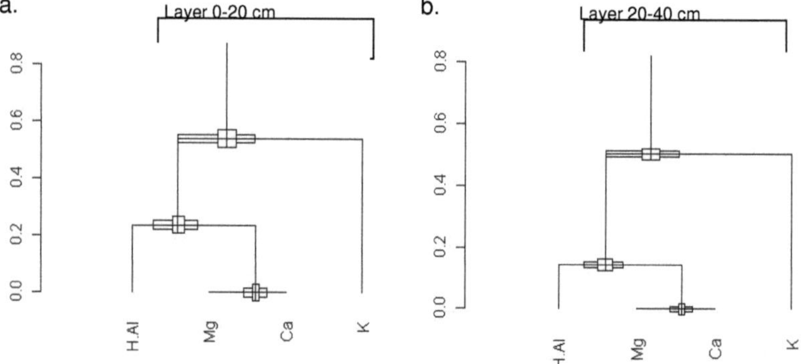

Figure 9. Dendograms of the compositional values in the 0-20 and 20-40 cm soil layer.

As presented in the ternary diagram, the distribution of the soil cation balances between the 0-20 and 20-40 cm layers overlapped (Figure 10). However, the small ellipses (Figure 11), which represent the confidence region of the averages, differed significantly between the 0-20 and 20-40 cm soil layers. The coordinate [K | Ca, Mg, H + Al] was higher in the 0-20 cm soil layer compared to the 20-40 cm layer, indicating that more K was accumulated in the surface layer, arising from the surface applications of potassium fertilizer.

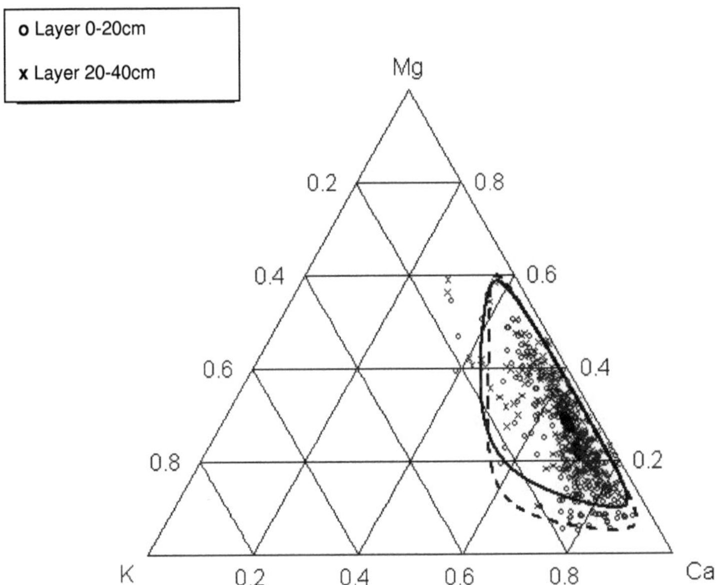

Figure 10. Ternary diagram of compositional values of K, Ca and Mg in soil layers 0-20 cm (dotted line) and 20-40 cm (continuous line), under N and K doses, with 95% confidence interval.

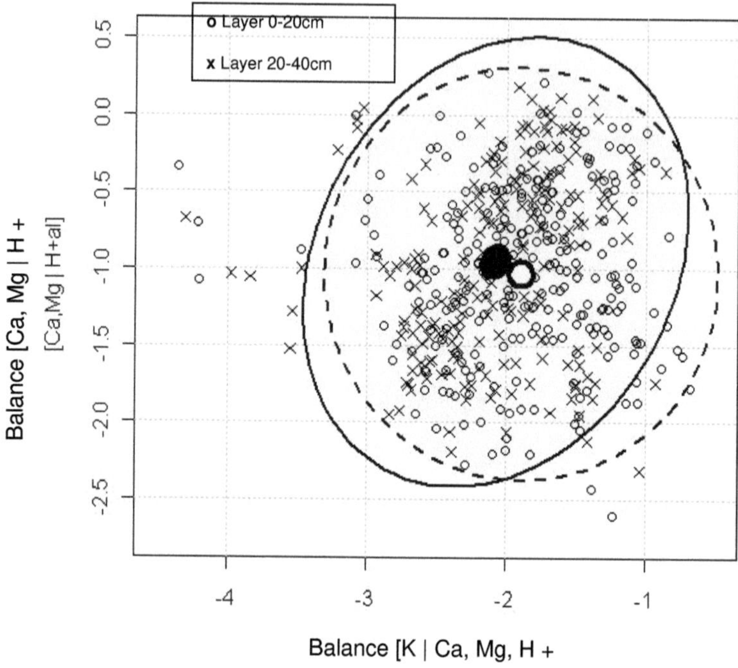

Figure 11. Relationship between the [Ca, Mg | H + Al] and [K | Ca, Mg, H + Al] balances in the 0-20 and 20-40 cm soil layers [the smaller ellipses are the average *ilr* coordinates for the 0-20 cm layer (white ellipses) and for the 20-40 cm layer (black ellipses)].

Critical soil cation balance

The average soil K concentration and balance [K | Ca, Mg, H + Al] were calculated between the beginning and end of the growing season as average soil conditions. As the fruit plant remains in the field throughout the year, and is continuously absorbing elements from the soil, soil analysis reflects the nutrient concentration at the time of sampling; as a result, for calculation purposes, only averages between samplings were considered. The concentration of K in the soil and the balance [K | Ca, Mg, H + Al] were related to fruit production and partitioned according to the Cate-Nelson procedure (Figures 12 and 13). In the second year of orchard establishment, there was no apparent relationship between fruit production and soil potassium concentration, or even between fruit production and the [K | Ca, Mg, H + Al] balance, due to the low yield of the guava plant, which was still in the formation phase.

Although difficult to define, the critical value of the concentration of K in the soil seems to be 1.60 mg dm-3 for 'Rica' (Figure 13a) and 1.20 mg dm-3 for 'Paluma' (Figure 13b) in the third year. The critical value of the balance [K | Ca, Mg, H + Al] was more clearly defined than the critical value of the concentration of K, determined by soil analysis, being -2.10 for the two cultivars (Figures 12a and 12b).

a.

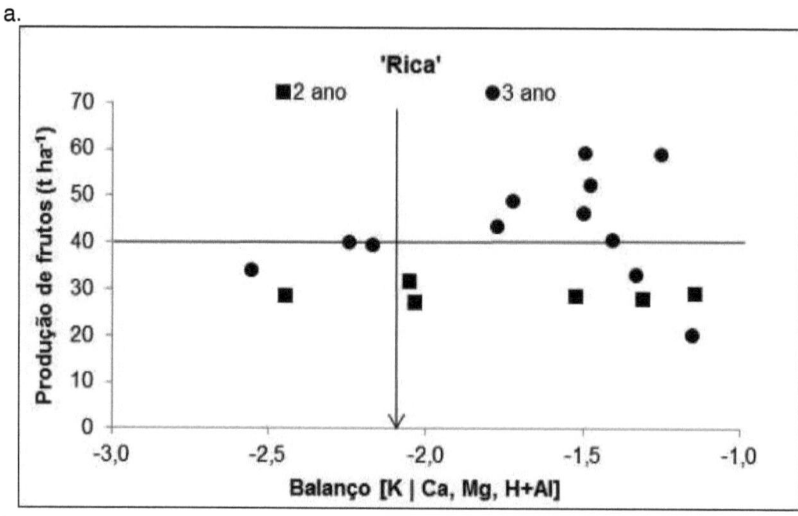

b.

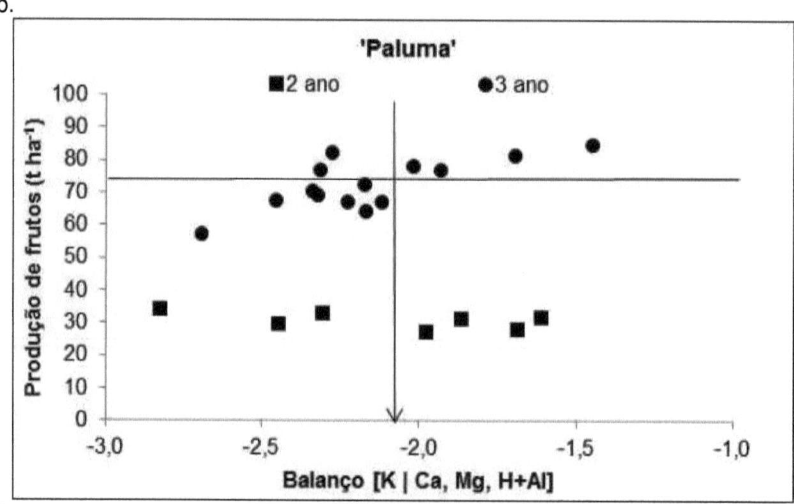

Figure 12. Cate-Nelson partitioning of the relationship between soil cation balance and guava production of the cultivars Rica (12a) and Paluma (12b), in experiments with doses of N and K, showing the critical value of the soil cation balance.

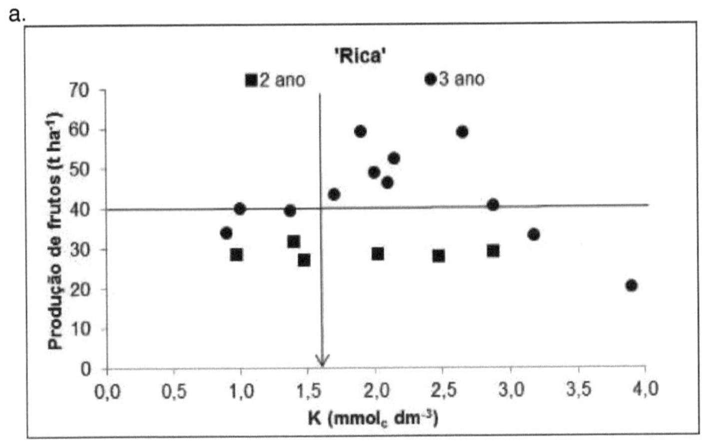

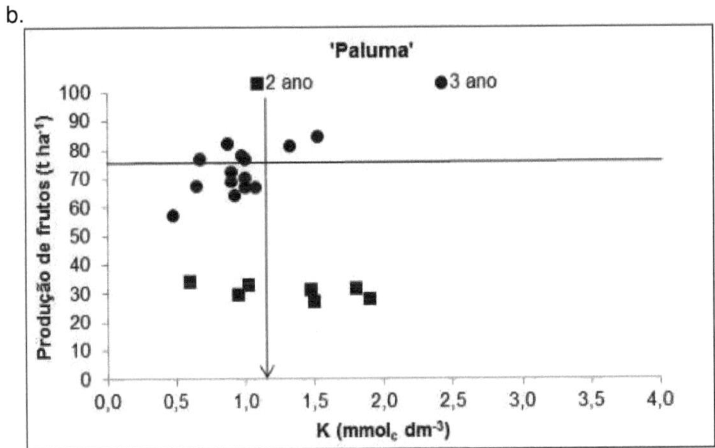

Figure 13. Cate-Nelson partitioning of the relationship between soil K concentration and yield of guava from Rica (13a) and Paluma (13b) cultivars in experiments with doses of N and K, showing the critical value of the soil balance.

There were nine samples in the true negative group (i.e. no imbalance) for soil K concentration (Figure 13), compared to eleven samples for the balance [K | Ca, Mg, H + Al] (Figure 12) for 'Rica' and 'Paluma'. Both indicators presented one sample in the false positive group (high yield classified as unbalanced) and two samples in the false negative group (low yield classified as balanced).

The success rate, calculated as the number of true positive and true negative points divided by the total number of points, was 85% for the soil K concentration and 92% for the balance [K | Ca, Mg, H + Al]. Therefore, the success rate of soil balance partitioning was higher compared to the success rate of soil K concentration, which indicates greater confidence in the results. The true positives are part of the high yielding subpopulation, targeted by growers and used for diagnostic purposes.

The critical values were approximately -1.28 for the [Ca, Mg | H+Al] balance (Figure 14) and 0.85 for the [Ca | Mg] balance (Figure 15). The critical values of - 2.10 for the [K | Ca, Mg, H + Al] balance, of -1.28 for the [Ca, Mg | H + Al] balance, and of 0.85 for the [Ca | Mg] balance were re-transformed to percent CTC saturation, corresponding to 2.0% K, 25% Ca, 7.5% Mg, and 65% (H+Al). For the Argissolo (CTC = 41 mmolc dm-3) and Latossolo (CTC = 54 mmolc dm-3) of the present trial, the critical level of K in the soil would be 0.80 and 1.10 mmolc K dm-3, respectively.

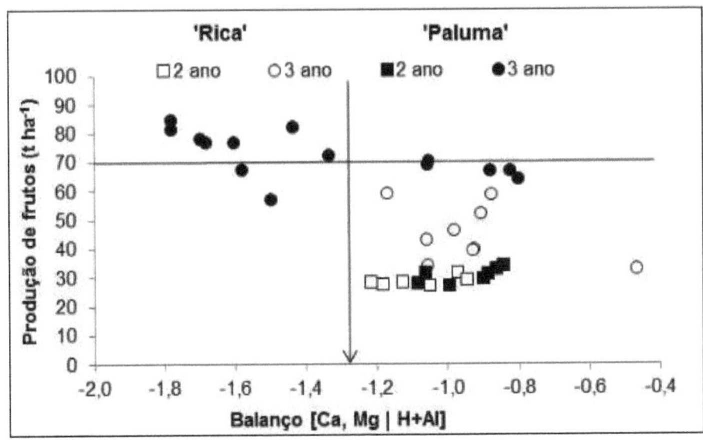

Figure 14. Cate-Nelson partitioning of the relationship between the soil cation balance [Ca, Mg | H + Al] and the production of guava from RicaPaluma cultivars, in experiments with doses of N and K, showing the critical value of the balance in the soil.

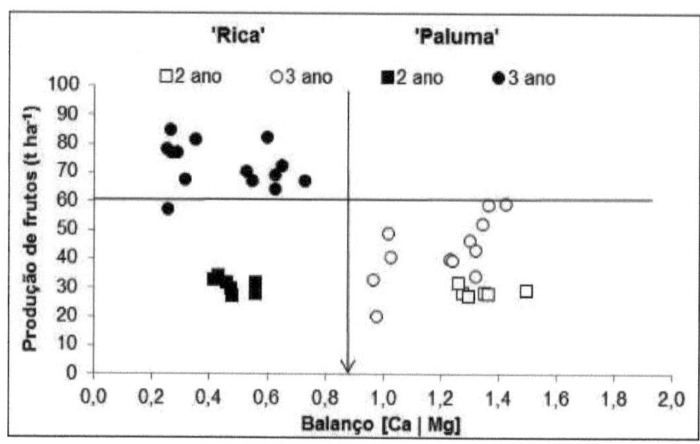

Figure 15. Cate-Nelson partitioning of the relationship between the soil cation balance [Ca | Mg] and the production of guava of the cultivars Rica and Paluma, in experiments with doses of N and K, showing the critical value of the balance in the soil.

To obtain a critical range of the balances, the ranges of balances from -1.90 to 2.10 for the [K | Ca, Mg, H + Al] balance were selected for both cultivars (Figure 12). For 'Paluma', the ranges of balances were from -1.78 to -0.80 for the [Ca, Mg | H + Al] balance (Figure 14), and from 0.27 to 0.73 for the [Ca | Mg] balance (Figure 15). The ranges of the Rica cultivar balances were -1.17 to -0.88 for the [Ca, Mg | H + Al] balance (Figure 14), and 0.97 to 1.42 for the [Ca | Mg] balance (Figure 15). Depending on the level of the other cation species, the K saturation ranges at the high yield level were 1.6 to 2.4% for the Paluma cultivar and 1.9 to 2.3% for the Rica cultivar. Ca saturation ranged from 11 to 33% for 'Paluma' and from 36 to 44% for 'Rica'. The Mg saturation ranged from 5 to 18% for 'Paluma' cultivar and from 5 to 9% for 'Rica' cultivar, while the H + Al saturation ranges were 53 to 80% for 'Paluma' and 48 to 61% for 'Rica'. However, these ranges still need to be validated using a wider database.

Potassium recommendation template

The soil balance [K | Ca, Mg, H + Al] was considered a soil fertility index for accounting for K under acid-base soil conditions, i.e. the geometric mean of [Ca, Mg, H + Al]. If the soil balance [K | Ca, Mg, H + Al] is too low, this means that either the level of K in the soil is too low, or, the acid-base balance that regulates the pH of the soil is too high. The soil pH is corrected by liming to restore the soil [Ca, Mg | H + Al] and [Ca | Mg] balances. At the critical [Ca, Mg | H + Al] balance value of -1.28, the critical pH (CaCl2) value was considered to be 4.5, determined from its close relationship with the soil balance values (Figure 16).

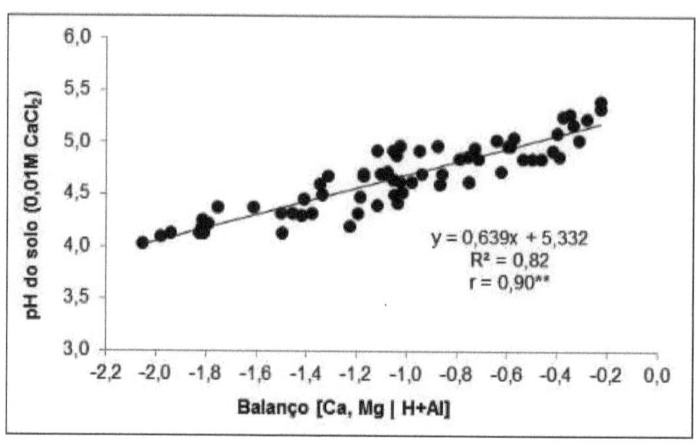

Figure 16. Relationship between the balance [Ca, Mg | H + Al] and soil pH.

Guava is acid tolerant, but adapts to a pH in H2O ranging from 4.5 to 7.0 (CRANE; BALERDI, 2009); however, it develops better in conditions of pH 5.0 to 6.0, according to Bourdelles and Estanove (1967). High fruit yields were obtained in a range of pH values in CaCl2 varying between 4.1 and 5.4 at both sites (Figures 5a and 6a). Below pH values of 5.4, the buffer capacity of Argissols and Latosols is mainly due to exchangeable Al (KAMPRATH, 1970). At pH values in H2O < 5.5 generally Al toxicity occurs in plants in Argysols and Latosols (KAMPRATH, 1984) and, as this did not occur in the present assay, guava appeared to b e   relatively insed to Al and Mn toxicity. Therefore, the objective of applying lime to acidic, highly fertilized soils in guava orchards would be to re-establish the cationic balance of the soil favourable for the high productivity of this fruit. The potassium fertilization model presented here is an approach for the "building" and maintenance of soil fertility, with the critical lower limit as reference. Nutrient exports and additions must be balanced to keep all three soil cation balances above these critical values. Since nutrient uptake increases with production and fertilizers containing N-

NH4 acidify the soil, potassic fertilizers, gypsum and liming materials containing Ca and Mg should be provided to keep the cation balances in the soil above critical lower limits.

The critical balance [K | Ca, Mg, H + Al] of -2.10 can be maintained in the 0-20 cm soil layer, by applying 109 g K2O per plant, in Paluma cultivar orchards and, 49 g K2O per plant in Rica cultivar orchards in the first year, doubling and tripling these doses in the second and third years respectively (Figure 17).

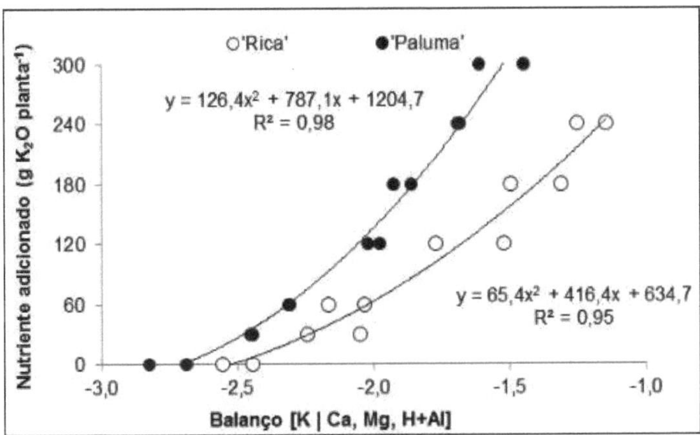

Figure 17. Relationship between the balance [K | Ca, Mg, H + Al] and the addition of K to the soil. [†] K doses in 1° year: zero, 30, 60, 120, 180, 240 and 300 g K2O plant-1; in 2° and 3° years, double and triple the initial doses respectively.

Although a satisfactory K dose cannot be defined from the yield response curve (Figure 2), K doses of 49 or 109 g K2O per plant in the first year, and double and triple in the second and third years, respectively, were supported by a significant contrast between the control (zero dose) and the fertilized treatments (Tables 6 and 7). In comparison, K removal by the crop was about 120 g K2O per plant in the first year, and double and triple in the second and third years, respectively (Figures 7 and 8). The application of 120 g K2O per plant in the first year and double and triple in the second and third years, respectively, is necessary to maintain the soil balance [K | Ca, Mg, H + Al] at -2.10 in the 20-40 cm layer (Figures 3a and 4a), also quite exploited by the roots of the guava trees.

There is concern regarding the application of TSBC methods in soils with low CTC, such as Latosols and Argissols, which can lead to mineral deficiency, since there are no absolute minimum levels defined. In soils with very low CEC, the quantity of a given element, although in a correct proportion in relation to the other elements, may prove to be too low to meet the needs of the crop. On the other hand, applying TSBC methods in soils with high CTC may result in excessive application of potassium fertilizers (KOPITTKE; MENZIES, 2007). In the case of guava orchards grown on two soils of low CTC in Brazil (40 - 50 mmolc dm-3), maintaining the balance [K | Ca, Mg, H + Al] above -2.10, with the use of potassium fertilizers, can guarantee high fruit production.

Conclusion

The soil cation balances in the 'Rica' and 'Paluma' guava orchards were arranged orthogonally to avoid numerical distortions and to facilitate the management of N and K fertilization in Latosols and Argissolos. The maintenance of the balance [K | Ca, Mg, H + Al] requires that K exports by the crop are compensated by equivalent amounts of potassium fertilizers. The *ilr* provided a reliable and unbiased index of the soil cation balance. Critical values of CTC saturation were determined to be 2.0% K, 25% Ca, 7.5% Mg and 65% (H + Al), with the critical pH (CaCl2) value equal to 4.5 for yields > 77 t ha-1.

References

AITCHISON, J. **The statistical analysis of compositional data.** London: Chapman & Hall, 1986. 416 p.

ANJANEYULU, K.; RAGHUPATHI, H. B. Identification of yield-limiting nutrients through DRIS leaf nutrient norms and indices in guava (*Psidium guajava*). **Indian Journal of Agricultural Sciences**, New Delhi, v. 79, p. 418-421, 2009.

ARORA, J. S.; SINGH, J. R. Effect of nitrogen, phosphorus and potassium sprays on guava (*Psidium guajava* L.). **Journal of the Japanese Society for Horticultural Science**, Kyoto, v. 39, p. 55-62, 1970.

BOURDELLES, J.; ESTANOVE, P. La goyave aux. Antilles. **Fruits**, Paris, v. 22, p. 397-412, 1967.

BUCCIANTI, A. Natural laws governing the distribution of the elements in geochemistry: the role of the log-ratio approach. In: PAWLOWSKY-GLAHN, V.; BUCCIANTI, A. (Eds). **Compositional data analysis:** theory and applications. New York: John Wiley and Sons, 2011. p. 255-266.

CHAYES, F. On correlation between variables of constant sum. **Journal of Geophysical Research**, Washington, v. 65, p. 4185-4193, 1960.

CRANE, J. H.; BALERDI, C. F. **Guava growing in the Florida home landscape.** Horticultural Sciences Department, Florida Cooperative Extension Service, Institute of Food and Agricultural Sciences, University of Florida, Florida, 2009. 8 p.

DIAZ-ZORITA, M.; PERFECT, E.; GROVE, J. H. Disruptive methods for assessing soil structure. **Soil and Tillage Research**, Amsterdam, v. 64, p. 3-22, 2002.

EGOZCUE, J. J.; PAWLOWSKY-GLAHN, V. Groups of parts and their balances in compositional data analysis. **Mathematical Geology**, New York, v.37, p. 795-828, 2005.

EGOZCUE, J. J.; PAWLOWSKY-GLAHN, V.; MATEU-FIGUERAS, G.; BARCELÓ-VIDAL, C. Isometric log-ratio transformations for compositional data analysis. **Mathematical Geology**, New York, v. 35, p. 279-300, 2003.

EMBRAPA. **Sistema Brasileiro de classificação de solos.** 2nd ed. Rio de Janeiro, Brazil: Embrapa Solos, 2006. 306 p.

FILZMOSER, P.; HRON, K. Robust statistical analysis. In: PAWLOWSKY-GLAHN, V.; BUCCIANTI, A. (Eds.). **Compostional data analysis**: theory and applications, New York: John Wiley and Sons, 2011. p. 57-72.

FRACARO, A. A.; PEREIRA, F. M. Distribuição do sistema radicular da goiabeira 'Rica' produzida a partir de estaquia herbácea. **Revista Brasileira de Fruticultura**, Jaboticabal, v. 26, p. 183-185, 2004.

GERALDSON, C. M. Factors affecting calcium nutrition of celery, tomato and pepper.
Soil Science Society of America Proceedings, Madison, v. 21, p. 621-625, 1957.

GERALDSON, C. M. Intensity and balance concept as an approach to optimal vegetable production. **Communications in Soil Science and Plant Analysis**, New York, v. 1, p. 187-196, 1970.

GERALDSON, C. M. Nutrient intensity and balance. In: STELLY, M. (Ed.). **Soil testing:** correlating and interpreting analytical results. Madison: American Society of Agronomy, 1984. p. 75-84.

GLYNN, P. E.; WEISBACH, P. C. **Clinical predictions rules:** a physical therapy reference manual. Sudbury: Jones and Barlett. Disponível em:
<   http://samples.jbpub.com/9780763775186/75186_FMXX_Glynh.pdf>.
Accessed on: 30 Jul. 2012.

IHAKA, R.; GENTLEMAN, R. R: A language for data analysis and graphics. **Journal of Computational and Graphical Statistics**, Alexandria, v. 5, p. 299-314, 1996.

KAMPRATH, E. J. Crop response to lime on soils in the tropics. In: ADAMS, F. (Ed.). **Soil acidity and liming.** 2nd. ed. Madison: American Society of Agronomy, 1984. p. 349-368.

KAMPRATH, E. J. Exchangeable aluminium as a criterion for liming leached mineral soils. **Soil Science Society of America Proceedings**, Madison, v. 34, p. 252-254, 1970.

KOPITTKE, P. M.; MENZIES, N. W. A review of the use of the basic cation saturation ratio and the "ideal" soil. **Soil Science Society of America Journal**, Wiscosin, v. 71, p. 259-265, 2007.

LEVY, G. J.; van der WATT, H. V. H.; SHAINBERG, I.; du PLESSIS, H. M. Potassium-calcium and sodium-calcium exchange on kaolinite and kaolinitic soils. **Soil Science Society of America Journal**, Madison, v. 52, p. 1259-1264, 1988.

LIEBHARDT, W. C. The basic cation saturation ratio concept and lime and potassium recommendations on Delaware's Coastal Plain soils. **Soil Science Society of America Journal**, Madison, v. 45, p. 544-549, 1981.

LOPEZ, J.; PARENT, L. E.; TREMBLAY, N.; GOSSELIN, A. Sulfate accumulation and Ca balance in hydroponic tomato culture. **Journal of Plant Nutrition**, Athens, v. 25, p. 1585-1597, 2002.

MA, C.; EGGLETON, R. A. Cation exchange capacity of kaolinite. **Clays and Clay Minerals**, Chantilly, v. 47, p. 174-180, 1999.

MALAVOLTA, E. **ABC da Adubação.** 5ª. ed. São Paulo: Agronômica Ceres Ltda, 1989. 292 p.

MALAVOLTA, E. **Manual de nutrição de plantas.** São Paulo, Brazil: Ceres, 2006. 638 p.

McLEAN, E. O. Contrasting concepts in soil test interpretation: sufficiency levels of available nutrients versus basic cation saturation ratios. In: STELLY, M. (Ed.). **Soil testing:** correlating and interpreting the analytical results. Madison: American Society of Agronomy, 1984. p. 39-54.

McLEAN, E. O.; HARTWIG, R. C.; ECKERT, D. J.; TRIPLETT, G. B. Basic cation saturation ratios as a basis for fertilizing and liming agronomic crops. II. Field studies.
Agronomy Journal, Madison, v. 75, p. 635-639, 1983.

MEHLICH, A. Uniformity of soil test results as influenced by volume weight. **Communications in Soil Science and Plant Analysis**, New York, v. 4, p. 475-486, 1973.

NATALE, W. **Diagnose da nutrição nitrogenada e potássica em duas cultivares de goiabeira (*Psidium guajava* L.), durante três anos.** 1993. 150 f. Thesis (PhD in Soil and Plant Nutrition) - Escola Superior de Agricultura "Luiz de Queiroz", Universidade de São Paulo, Piracicaba, 1993.

NATALE, W. **Resposta da goiabeira à adubação fosfatada.** 1999. 132 p. Thesis - Faculdade de Ciências Agrárias e Veterinárias - Unesp, Jaboticabal, 1999.

NATALE, W.; COUTINHO, E. L. M.; BOARETTO, A. E.; BANZATTO, D. A. Nutrient foliar content for high productivity cultivars of guava in Brazil. In: TAGLIAVINI, M. (Ed.). Proceedings ISHS on Foliar Nutrition, **Acta Horticulturae**, Bologna, v. 594, p. 383- 386, 2002.

NELSON, L. A.; ANDERSON, R. L. Partitioning of soil test-crop response probability. In: STELLY, M. (Ed.). **Soil testing:** Correlating and interpreting the analytical results. Madison: American Society of Agronomy, 1984. p. 19-38.

OSMAN, S. M.; ABD EL-RAHMAN, A. E. M. Effect of slow release nitrogen fertilization on growth and fruiting of guava under Mid Sinai conditions. **Australian Journal of Basic and Applied Science**, Pakistan, v. 3, p. 4366-4375, 2009.

PARENT, L. E. Diagnosis of the nutrient compositional space of fruit crops. **Revista Brasileira de Fruticultura**, Jaboticabal, v. 33, p. 321-334, 2011.

PARENT, L. E.; CISSÉ, S.; TREMBLAY, N.; BÉLAIR, G. Row-centred log ratios as nutrient indexes for saturated extracts of organic soils. **Canadian Journal of Soil Science**, Ottawa, v. 77, p. 571-578, 1997.

PARENT, L. E.; DAFIR, M. A theoretical concept of compositional nutrient diagnosis. **Journal of the American Society for Horticultural Science**, Mount Vernon, v.117, p.239-242, 1992.

PARENT, L. E.; de ALMEIDA, C. X.; PARENT, S. E.; HERNANDES, A.; EGOZCUE,
J. J.; KÄTTERER, T.; GÜLSER, C.; BOLINDER, M. A.; ANDRÉN, O.; ANCTIL, F.; CENTURION, J. F.; NATALE, W. Compositional analysis for an unbiased measure of soil aggregation. **Geoderma**, Amsterdam, v. 179, p. 123-131, 2012.

PARENT, L. E.; NATALE, W.; ZIADI, N. Compositional nutrient diagnosis of corn using the Mahalanobis distance as nutrient imbalance index. **Canadian Journal of Soil Science**, Ottawa, v. 89, p. 383-390, 2009.

PARENT, S. E.; KARAM, A.; PARENT, L. E. Compositional modeling of C mineralization of organic materials in soils. In: INTERNATIONAL SYMPOSIUM OF AGRICULTURAL WASTE MANAGEMENT, 2., 2011, Foz do Iguaçu. **Proceedings...** , Foz do Iguaçu, Brazil, 2011.

PEARSON, K. Mathematical contributions to the theory of evolution - On a form of spurious correlation which may arise when indices are used in the measurement of organs. **Proceedings of the Royal Society of London**, London, v. LX, p. 489-502, 1897.

PEREIRA, F. M. 'Rica' and 'Paluma': new guava cultivars. In: CONGRESSO BRASILEIRO DE FRUTICULTURA, 7, 1983, Florianópolis. **Proceedings...** Florianópolis: Brazilian Society of Fruitculture/EMPASC, 1984. p. 524-528.

QUAGGIO, J. A.; van RAIJ, B.; MALAVOLTA, E. Alternative use of the SMP-buffer solution to determine lime requirement of soils. **Communications in Soil Science and Plant Analysis**, New York, v. 16, p. 245-260, 1985.

RAIJ, B. van; QUAGGIO, J. A.; CANTARELLA, H.; FERREIRA, M. E.; LOPES, A. S.;
BATAGLIA, O. C. **Análise química do solo para fins de fertilidade.** Campinas: Cargill Foundation, 1987. 170 p.

SAS INSTITUTE. **SAS/STAT**: user's Guide. Version 9.2. Cary: SAS Institute, 2009. 7869 p.

SHOEMAKER, H. E.; McLEAN, E. O.; PRATT, P. F. Buffer methods for determining lime requirement of soils with appreciable amounts of extractable aluminium. **Soil Science Society of America Proceedings**, Madison, v. 25, p. 274-277, 1961.

TANNER, J. Fallacy of per-weight and per-surface area standards, and their relation to spurious correlation. **Journal of Applied Physiology**, Bethesda, v. 2, p. 1-15, 1949.

van den BOOGART, K. G.; TOLOSANA-DELGADO, R. "Compositions": a unified R package to analyze compositional data. **Computers & Geosciences, [S.l.]**, v. 34, p. 320-338, 2008.

WELTJE, G. J. Quantitative analysis of dentrial modes: statistically rigorous confidence regions in ternary diagrams and their use in sedimentary petrology. **Earth Science Reviews**, [S.l.], v. 57, p. 211-253, 2002.

CHAPTER 3 - Nutritional balance of guava trees based on fertilisation and liming

The response of guava to fertilization and liming can be monitored by plant tissue analysis. The nutrient profile of each crop, used for the diagnosis of plant nutrition, is defined in relation to standard nutrient contents. However, the standard nutrient contents are constantly criticized for not considering the interactions that occur between elements and, for generating numerical trends, due to data redundancy, scale dependence and non-normal distribution. Compositional data analysis techniques can control these biased data by balancing the groups of nutrients, such as those involved in fertilization and liming. The use of orthonormal log isometric ratios (*ilr*), sequentially arranged, avoid numerical biases inherent in compositional data. Thus, the objective of the work was to evaluate a system of isometric log ratios (*ilr*) to correct such distortions, relating the nutrient balances of plant tissues, with guava production, in fertilized and limed 'Paluma' orchards, in order to adjust the current nutrient patterns to the equilibrium range of the most productive guava trees. Data from experiments conducted for three years with N, P and K and, for seven years, data from an experiment with lime doses, in guava orchards on Latosols, were used. The foliar contents of N, P, K, Ca and Mg were evaluated annually. The foliar balances [N, P, K | Ca, Mg], [N, P | K], [N | P] and [Ca | Mg] were selected to separate the effects of fertilizers (N-K) and liming (Ca-Mg) on nutrient balance. The foliar balances were more influenced by liming than by fertilization. The productivity of guava trees and their foliar nutrient balance allowed the definition of nutrient balance ranges and their validation with the critical content ranges currently used in Brazil, as well as their combination in *ilr* coordinates.

Keywords: compositional data analysis, foliar nutrient balance, Aitchison geometry, plant mineral nutrition, *Psidium guajava* L., isometric log ratios (*ilr*)

Introduction

Brazil is the world's largest producer of red guava, a plant native to Tropical America. The guava is a fruit consumed *fresh* or processed into jelly and juice. The Paluma cultivar is precocious, vigorous and productive (PEREIRA, 1984). Its fruits are oval, tasty, with an average weight between 140 and 250 g, yellow skin and red pulp, firm, resistant to transport and storage, and suitable both for processing and for the fresh fruit market.

Guava (Psidium guajava L.) is responsive to fertilization (ARORA; SINGH, 1970; NATALE et al., 1996; NATALE et al., 2002a; ANJANEYULU; RAGHUPATHI,
2009; OSMAN; ABD EL-RAHMAN, 2009) and liming (PRADO; NATALE, 2004, 2008;
NATALE et al., 2005; NATALE et al., 2007). To achieve a nutritional balance, fertilization and liming should be balanced. The response of the guava plant to fertilization and liming can be monitored by plant tissue analysis. Leaf analysis has an advantage over soil analysis as a diagnostic tool for fruit crops, which have a deep root system, because the plant has access to more nutrients in the soil at depth than would be determined by normal soil analysis procedures (SMITH et al., 1985) and is gaining in popularity (NATALE et al., 2002a).

The interpretation of plant tissue analyses is based on the premise that alterations in the supply of nutrients are reflected in the contents of these elements in plant tissue, being accentuated in certain phases of plant development (BOULD, 1968). However, the standard nutrient contents have been criticized for not considering the interactions between nutrients (BATES, 1971), since several double and multiple interactions have been well documented in plants (BERGMANN, 1988; MALAVOLTA, 2006). The Integrated Diagnosis and Recommendation System (DRIS) uses dual nutrient relationships and can partly explain interactions between elements (WALWORTH; SUMNER, 1988); however, it is geometrically deficient, as demonstrated in Parent (2011). Parent and Dafir (1992) adjusted the DRIS procedure so that it allowed the transformation of the centred log ratio, used to analyse compositional data such as nutrient contents (AITCHISON, 1986).

Plant tissue data convey relative information, being intrinsically multivariate, since each component cannot be interpreted in isolation, without relating it to the other components (TOLONASA- DELGADO; VAN DEN BOOGART, 2011). Lagatu and Maume (1935) were the first to illustrate the effects of changing proportions on nutrient interaction in a closed system, using a ternary diagram representing a three-component mixing system that closes at 100%. As the compositional vector with D-parts contains one redundant part, calculated by the difference between the total value and the value of the other parts, there is then at least one negative correlation (AITCHISON, 1986) and D-1 non-redundant variables (EGOZCUE; PAWLOWSKY-GLAHN, 2005). Furthermore, there are three dual relations in a composition with D-parts, however, one is redundant since it is calculated from the other two relations, i.e. $X/Z = [(X/Y) / (Z/Y)]$.

To avoid numerical bias and redundancy inherent in compositional data and generate D-1 variables, Egozcue and Pawlowski-Glahn (2005) proposed the use of sequentially arranged, orthonormal log isometric (*ilr*) ratios based on the principle of orthogonality. The *ilr* coordinates can be projected into Euclidean space, a geometric structure that allows statistical calculations without numerical bias. The *ilr* method is performed in three steps: the representation of data in *ilr* coordinates; the analysis of variance of the coordinates as real random variables; and, the interpretation of the results in terms of balances (EGOZCUE; PAWLOWSKY-GLAHN, 2011).

Thus, the hypothesis of the present study is that the interpretation of plant tissue analyses using the *ilr* concept is better than the interpretations coming from the use of other approaches. Thus, the objective of this work was to compare the effects of NPK fertilization and liming on the foliar balances of macronutrients (N, P, K, Ca and Mg) of guava trees of the Paluma cultivar, using the unbiased *ilr* concept and the standard nutrients currently used in Brazil.

Material and methods

The results used to form the database for this study were from experiments conducted with the guava crop by Natale (1993) and Natale (1999), collected during the formation and development of the guava trees, totaling 468 observations; and from an experiment with liming doses conducted by Natale et al. (2007) and Prado and Natale (2008) with the Paluma cultivar, for seven years (1999 - 2006), on a typical red dystrophic medium textured Latosol (EMBRAPA, 2006), in Bebedouro (São Paulo), totaling 100 observations.

Fertilisation experiments

Three experiments with NPK doses were conducted during three consecutive agricultural seasons in São Carlos, São Paulo State, Brazil, on a red-yellow, epieutrophic, endodistrophic, moderate A, medium texture, cerrado phase and flat relief (EMBRAPA, 2006). The experiments with NPK fertilization were initiated in April/May 1989, with one-year-old seedlings of the Paluma cultivar, propagated by herbaceous cuttings and conducted for three consecutive years. The plots were made up of four plants, of which the first was the border plant, in 7 x 5 m spacing, considering only 3 plants as useful for evaluation purposes. Three different areas received variable doses of N, P and K fertilizers. The randomized block design was used, with seven treatments and four repetitions. The fertilizer doses were increased proportionally each year, taking into account the removal of nutrients by the harvest and, principally, the increase in the demand for nutrients due to the development of the fruit trees.

In the first year, the treatments with nitrogen doses were zero, 30, 60, 120, 180, 240 and 300 g N plant-1, supplemented with 120 g P2O5 plant-1 and 120 g K2O plant-1. The initial N doses were doubled and tripled in the second and third years, respectively. The initial P and K doses were doubled in the second year, and in the third year, the P and K doses were 240 g P2O5 planta-1 and 360 g K2O planta-1, respectively.

The treatments with phosphorus doses, in the first year, were zero, 30, 60, 120, 180, 240 and 300 g P2O5 plant-1, supplemented with 120 g N plant-1 and 120 g K2O plant-1. In the second and third years, the initial P doses were doubled, while the N and K doses were doubled and tripled respectively.

In the first year, the treatments with potassium doses were zero, 30, 60, 120, 180, 240 and 300 g K2O plant-1, supplemented with 120 g N plant-1 and 120 g P2O5 plant-1. In the second year, the doses of N, P and K were doubled and in the third year, the doses of N and K were tripled and supplemented with 240 g of P2O5 planta-1.

In the experiment with N doses, the sources of N, P and K used were, respectively, ammonium nitrate (34% N), simple superphosphate (18% P2O5) and potassium chloride (60% K2O). In the experiments with doses of P and K, the sources of N, P and K employed were ammonium sulphate (20% N), triple superphosphate (44% P2O5) and potassium chloride (60% K2O) respectively. Fertilizers were applied manually around the plant in a 40 cm wide band around the crown.

During the conduction of the experiments in the field, the normal cultural treatments necessary for the good development of the plants were carried out. The conduction system was of cleaning pruning and conduction, with one production per year, non-irrigated. To evaluate the production, the fruits were harvested from one to three times a week, which occurred from January/February to May/June of each year.

Liming experiment

The treatments consisted of five doses of lime and four repetitions, arranged in a randomized block design. The lime was incorporated into the soil in August 1999, and the guava seedlings of the Paluma cultivar, propagated by cuttings, were planted in December 1999. The treatments consisted of increasing doses of lime, taking as a reference the dose necessary to reach a base saturation (V) equal to 70%, considered ideal for guava (SANTOS; QUAGGIO, 1996). The doses of lime were calculated considering the results mean values of the chemical analysis of the 0-20 and 20-40 cm layers and also for incorporation in the 0-30 cm layer, as follows: D0 = zero limestone; D1 = half the dose to achieve V% of 70%; D2 = the dose to achieve V% of 70%; D3 = 1.5 times the dose to achieve V% of 70%; and, D4 = 2 times the dose to achieve V% of 70%, corresponding to zero; 1.9; 3.8; 5.6 and 7.4 tons of limestone ha-1 respectively. The limestone used contained 455.5 g CaO kg-1 and 102.1 g MgO kg-1 (molar ratio Ca/Mg = 3.2). The relative efficiency (lime granulometry distribution) was 93.72% and the relative power of total neutralization equal to 106.85%. Limestone was applied manually in a uniform manner over the whole area, half before incorporation with a mouldboard plough and the other half after this operation, but before incorporation with a plough harrow.

The plots were composed of five plants, spaced 7 x 4.2 m, with the three central guava trees considered useful for evaluation purposes. The conduction system was of cleaning pruning and conduction, with one production per year, non-irrigated. The fertilization during the whole experimental period was based on the recommendations of Natale et al. (1996), being 140 g N and 112 g K2O in the first year; 200 g N and 50 g K2O in the second year; 200 g N, 30 g P2O5 and 150 g K2O in the third year; 400 g N, 60 g P2O5 and 300 g K2O in the fourth year; 800 g N, 50 g P2O5 and 300 g K2O in the fifth year; and 800 g N, 50 g P2O5 and 300 g K2O in the sixth year. Fertilizers were applied 3 to 4 times during the growing season. In the second, third and fourth years, an organic bovine fertilizer was added, at a dose of 40 L plant-1. The fertilizer sources used were urea and ammonium sulphate (1:1), triple superphosphate or mono-ammonium phosphate and potassium chloride. Fertilizers were applied around the plant in a 40 cm wide band, in the projection of the crown of the guava trees. Between 2002 and 2006 guava production, number and weight of fruits were determined weekly and occasionally two or three times a week during the harvest period, which lasted from January/February to May/June of each year.

Sampling and chemical analysis of soil and leaves

For the initial characterization, the soils were sampled in four locations per plant (North, South, East and West), in the 0-20 cm soil layer at the beginning of the experiment, in order to compose the composite sample (12 points) per plot. The soil samples were air dried and analyzed for pH (CaCl2), organic matter, K, Ca, Mg and (H + Al), according to the methodology of Raij et al. (1987). The soil analyses are presented in Table 1.

For leaf analysis, thirty pairs of leaves with petioles, mature (the third from the tip of the branch), were sampled at flowering time, 1.5 m above the ground, as described by Natale et al. (1996). The tissue samples were dried in an oven at 65°C and analysed for N, P, K, Ca and Mg, according to Bataglia et al. (1983). N was determined by the micro Kjeldahl method. P was quantified by colorimetry and the cations by atomic absorption spectrophotometry, after nitric-perchloric digestion.

Table 1. Soil properties at the beginning of the experiments

Experiments	pH CaCl2	Matter Organic	P resin	K	Ca	Mg	H + Al
		g dm-3	mg dm-3	mmolc dm-3			
Fertilisation Red-Yellow Latosol	5,3	19	19	0,5	14	8	18
Liming Red Latosol	4,7	18	6	1,3	9	4	40

Foliar nutrient balance

The compositional vector of the tissue analyses was composed of the nutrients N, P, K, Ca and Mg. Sequential binary partitioning (PBS) can be arranged to facilitate the interpretation of the balances in relation to the objectives of this study. Parent (2011) initiated sequential binary partitioning (PBS) between nutrient groups by contrasting anions (N, P) with cations (K,

Ca, Mg) in a physiological perspective. In the present study, the balances were initiated with a contrast between fertilizer nutrients (N, P, K) and limestone nutrients (Ca, Mg) in order to

separate the effects of fertilization and liming on the foliar nutrient balance (Table 2).

Table 2. Orthogonal sequential partitioning of leaf macronutrients

Ilr	Nutrients					Balance	r s †	*ilr* calculation
	N	P	K	Ca	Mg			
1	1	1	1	-1	-1	N, P, K | Ca, Mg]	3 2	$\sqrt{\frac{3x2}{3}} \ln \left(\frac{(NxPxK)^{1/3}}{} \right) 3 \cdot$
2	1	1	-1	0	0	N, P | K]	2 1	$\sqrt{\frac{2x1}{}} \ln \left(\frac{(NxP)^{1/2}}{} \right) 2 +$
3	1	-1	0	0	0	[N | P]	1 1	$\sqrt{\frac{1x1}{1+1}} \ln \left(\frac{N}{P} \right) -$
4	0	0	0	1	-1	Ca | Mg]	1 1	$\sqrt{\frac{1x1}{1+1}} \ln \left(\frac{Ca}{Mg} \right) -$

†r = number of positive signs and s = number of negative signs.

Isometric log ratio transformation

Cadacoordenada *ilr* has been calculated as follows (EGOZCUE; PAWLOWSKY-GLAHN, 2005):

$$ilr_j = \sqrt{\frac{rs}{r+s}} \ln \frac{g(c+)}{g(c-)} \text{ with } j = [1, 2, \ldots, D - 1] \tag{7}$$

where *ilrj* is the isometric *ji-th* log ratio, r and s represent the numbers of components in positive (+) and negative (-) groups in each binary partition, respectively, $g(c+)$ is the geometric mean of the components in the positive groups c+, and $g(c-)$ is the geometric mean of the components in the negative groups c-

(Table

2). The $\sqrt{\dfrac{rs}{r+s}}$ coefficient represents the balance between the number of components in the positive and negative groups.

Calculations

The critical ranges of balances were obtained from standard nutrient contents, indicated by Natale et al. (2002a) and Maia et al. (2007). In summary, the lower limits of contents were established using the lowest values in the numerator and the highest in the denominator, while the upper limits were obtained by the ratio between the highest values of contents in the numerator and the lowest in the denominator. Natale et al. (2002b) proposed ranges of contents of 20-23 g N kg-1; 1.4-1.8 g P kg-1; 14-17 g K kg-1; 7-11 g Ca kg-1; and, 3.4-4.0 g Mg kg-1 during flowering. Maia et al. (2007) suggested ranges of contents of 20.2-25.3 g N kg-1; 1.4-1.5 g P kg-1; 19.0-21.7 g K kg-1; 7.7-8.3 g Ca kg-1; and 2.7-2.8 g Mg kg-1 in leaves collected between budbreak and fruit set.

Statistical analyses

The *ilr* calculations and discriminant analyses were conducted in the R statistical computing environment (IHAKA; GENTLEMAN, 1996), which is a free *software,* using the compositional package (VAN DEN BOOGART; TOLOSANA- DELGADO, 2008) for compositional data analysis. Analysis of variance was performed by PROC MIXED of SAS (*Statistical Analysis System,* version 9.2., 2009), and its significance was checked by the t-test.

Results and discussion

Fruit production

The productivity of guava increased rapidly during the first three years of crop development, reaching maximum production after that, declining thereafter, which may be due to alternation of production, which occurs in many fruit trees (ISAGI et al., 1997; MONSELISE; GOLDSCHMIDT, 1982), although there is no scientific report of such alternation in guava culture. Fruit production was higher in the fertilization experiments (Figure 1) than in the liming experiment (Figure 2), although they had the same conduction system. The response of the guava plants to the addition of N and K was significant only in the third year, with linear adjustment (Figures 1a and 1b), while the response to liming was significant in the first four years, except in the fifth year (2006), showing linear adjustment in the first, third and fourth years (2002, 2004 and 2005) and quadratic adjustment in the second year (2003) (Figure 2). The guava crop generally shows little response to P additions (Figure 1c) under Brazilian conditions (NATALE et al., 2002b).

The production increased annually until it reached its maximum value in the third year after crop establishment (Figure 1). This is why the doses of N, P and K were adjusted annually during the experiment, in order to account for exports through harvests and, principally, to supply the demand of the plant due to its development. The linear increase in yield with the addition of N (Figure 1a) was quite modest, and may be due to the competition between soil micro-organisms and the fruit for available N, related to a rapid recycling rate of soil organic matter under tropical conditions (KRULL, 2002) and associated with a decrease in the C/N ratio (ZHANG; HE, 2004). The experiments provided a broad spectrum of fruit production and foliar nutrient balances over the years.

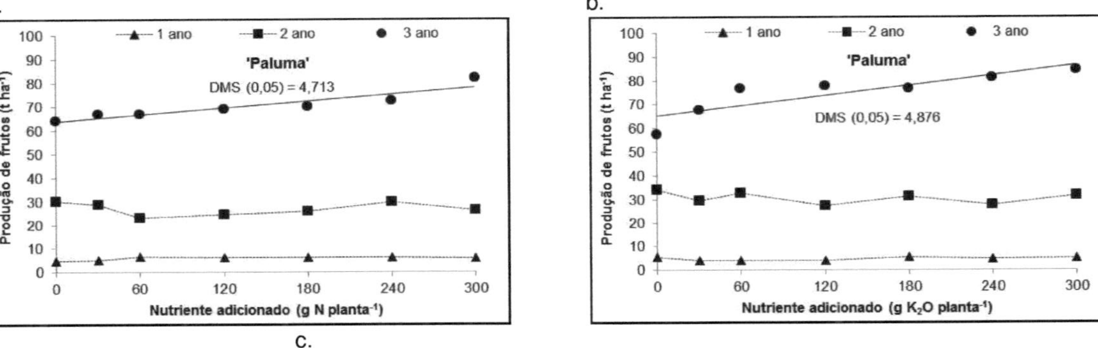
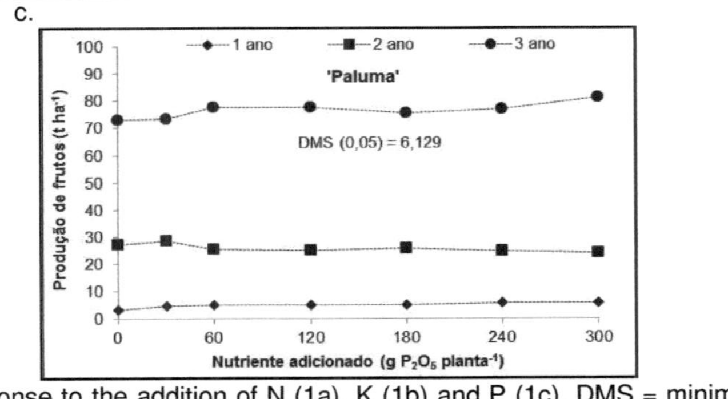

Figure 1. Guava yield in response to the addition of N (1a), K (1b) and P (1c). DMS = minimum significant difference at 0.05 probability. [†] N and K doses in the1° year: zero, 30, 60, 120, 180, 240 and 300g N or K2O plant-1 respectively, in the 2° and 3° years double and triple the initial doses respectively. P doses in the1° year: zero, 30, 60, 120, 180, 180, 240 and 300g of P2O5 planta-1, in the 2° and 3° years twice the initial doses.

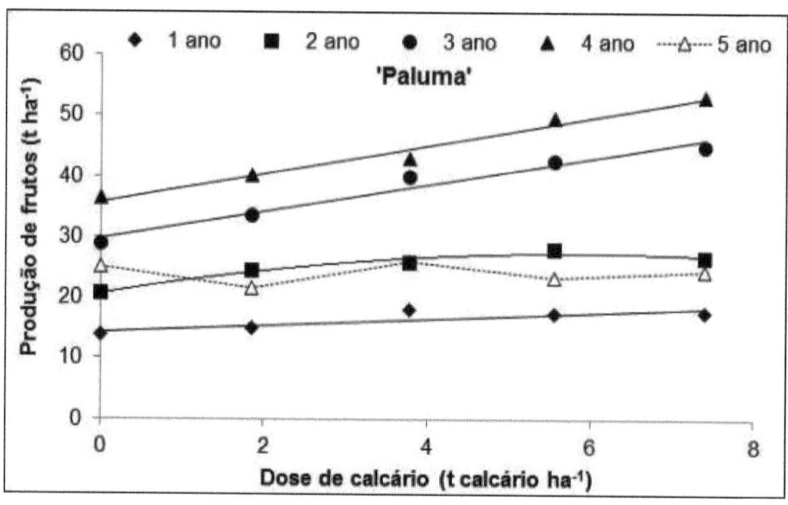

Figure 2. guava yield in response to the addition of lime.

Critical equilibrium intervals

The ranges of leaf nutrient balances obtained by combining the standard nutrient contents reported by Natale et al. (2002b) and Maia et al. (2007) are presented in Table 3. The ranges of foliar balances, calculated from the indications of Maia et al. (2007), were outside the range of the foliar balances of the higher yielding fruit trees in this study (Figure 3). In particular, the foliar [N, P, K | Ca, Mg] (Figure 3a), [N | P] (Figure 3c) and [Ca | Mg] (Figure 3d) balances were too high, driven by lower foliar Ca, Mg and P contents and higher foliar K contents. The patterns derived from Natale et al. (2002b) appear to be sufficient for the foliar balances [N, P, K | Ca, Mg] (Figure 3a) and [N,P|K] (Figure 3b), but could be adjusted for the foliar balances [N | P] (Figure 3c) and [Ca | Mg] (Figure 3d), as suggested in Table 3, in order to encompass data that also showed high yield.

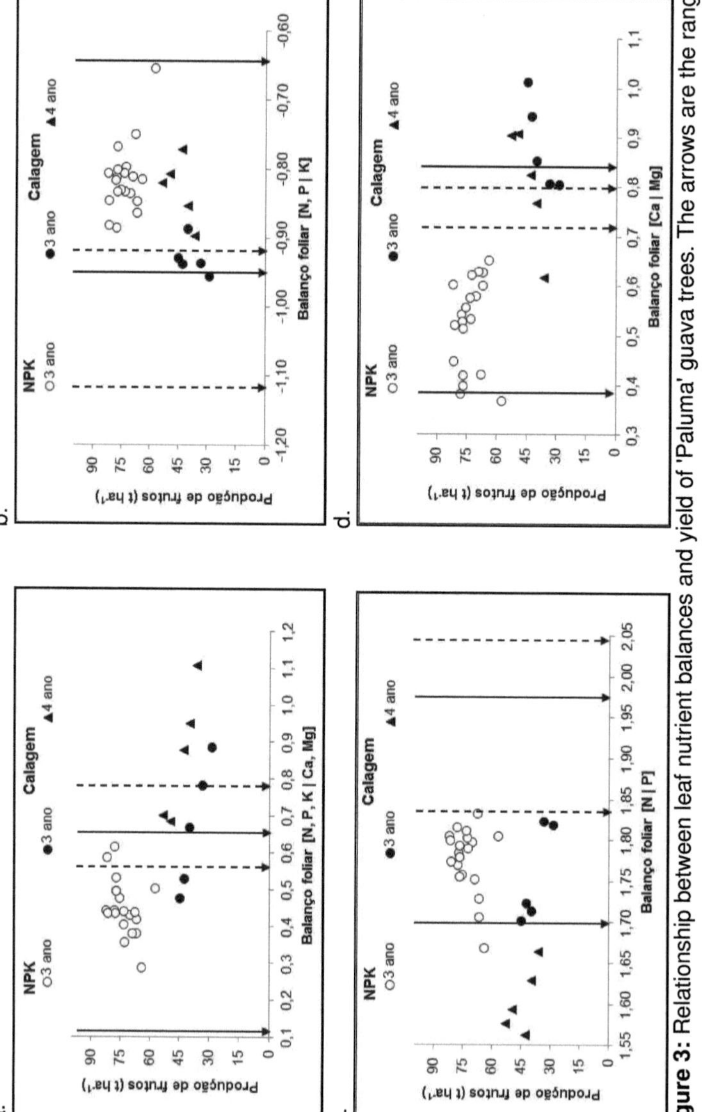

Figure 3: Relationship between leaf nutrient balances and yield of 'Paluma' guava trees. The arrows are the ranges of nutrient balances (LI and LS) calculated from Natale (2002b) (filled arrows) and Maia et al. (2007) (dashed arrows).

Table 3 - Best ranges of nutrient balances according to the nutrient sufficiency ranges described by Natale et al. (2002a) and Maia et al. (2007)

Leaf Balance	Range of Foliar Nutrient Balances					
	NATALE et al. (2002a)		MAIA et al. (2007)		NATALE et al. (2002a) Adjusted	
	LI	LS	LI	LS	LI	LS
N, P, K | Ca, Mg] N, P | K]	0,108 - 0,953	0,658 - 0,635	0,572 - 1,148	0,789 - 0,920	0,108 - 0,953	0,658 - 0,635
[N | P]	1,703	1,979	1,839	2,047	1,65	1,85
Ca | Mg]	0,396	0,830	0,715	0,794	0,35	0,65

† LI = lower limit and LS = upper limit.

Seasonal variations in foliar nutrient balances are presented in Figures 4 to 7. Foliar N balances do not appear to be limiting (Figure 4), despite the significant response of fruit yield to N addition in the third year (Figure 1a). Similarly, phosphorus was not deficient (Figure 5). Potassium started to be limiting, as shown by the excessively high leaf balance [N, P | K] (Figure 6). Guava is, in general, sensitive to K fertilization (NATALE et al., 1996), as the crop extracts 1,554 g K per ton of fresh fruit (NATALE et al., 2002b). In comparison, it has been reported that other guava cultivars can extract from 726 to 3130 g K per tonne of fresh fruit (HAAG et al, 1993;

ATALE et al, 2002b). Thus, 80 t ha-1 of Paluma cultivar fruits extract 124 kg K ha-1 or, approximately 150 kg of K2O ha-1. Since the response of the crop and the foliar nutrient balance were satisfactory with K applications of 30 to 60 g K2O plant-1, the K reserves in the soil seem to largely support the K requirements of the crop at high yield level. Natale et al. (2001) showed that feldspar in the silt fraction and micas in the clay fraction contributed more to the total soil K concentration in a Brazilian Latosol under guava production. The high surface charge density of kaolinite results in its high affinity for potassium (LEVY et al., 1988).

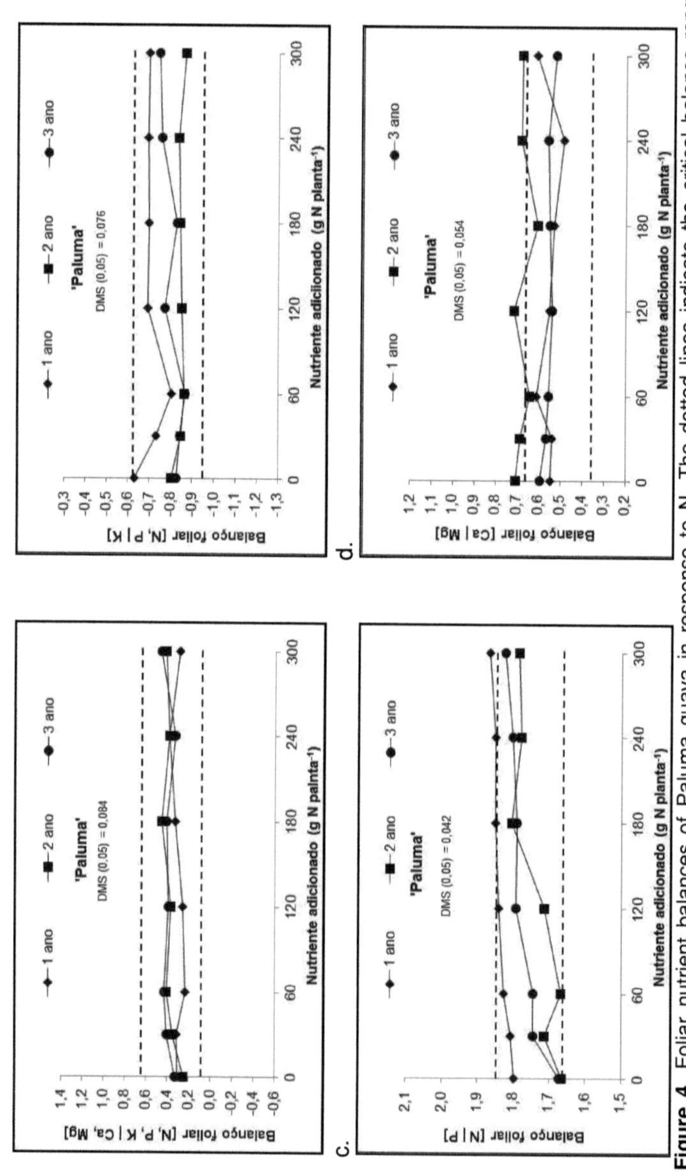

Figure 4. Foliar nutrient balances of Paluma guava in response to N. The dotted lines indicate the critical balance range. DMS = minimum significant difference at 0.05 probability. † N doses in the 1° year: zero, 30, 60, 120, 180, 240 and 300g of N plant-1, in the 2° and 3° years double and triple the initial doses, respectively.

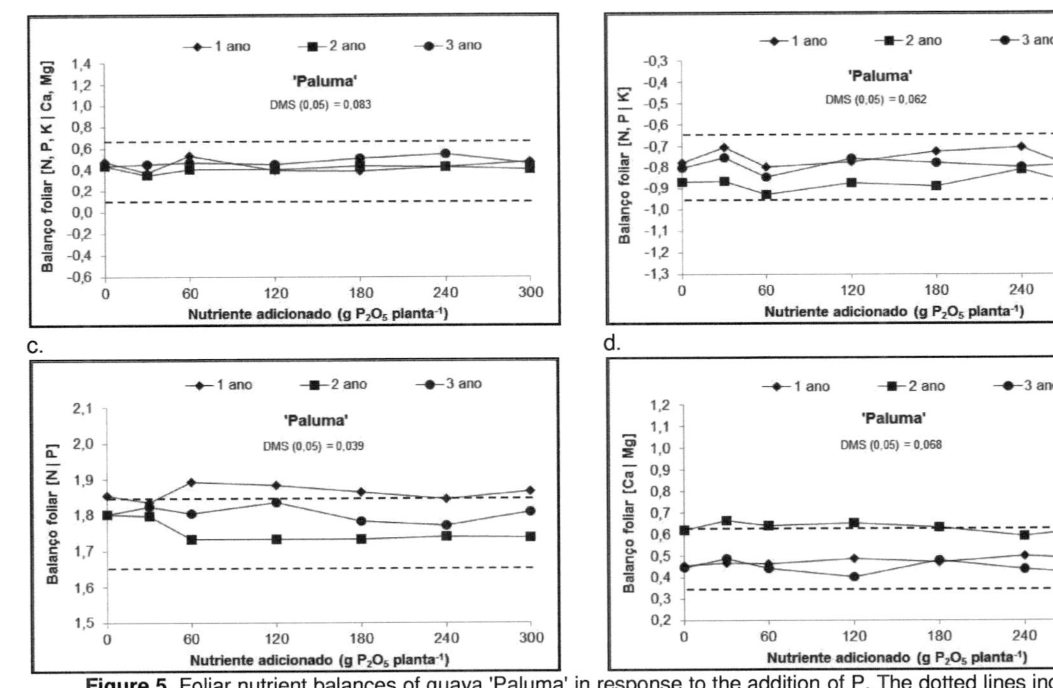

Figure 5. Foliar nutrient balances of guava 'Paluma' in response to the addition of P. The dotted lines indicate the critical balance range. DMS = minimum significant difference at 0.05 probability. [†] P doses in the1° year: zero, 30, 60, 120, 180, 240 and 300g of P2O5 planta-1, in the 2° and 3° years twice the initial doses.

107

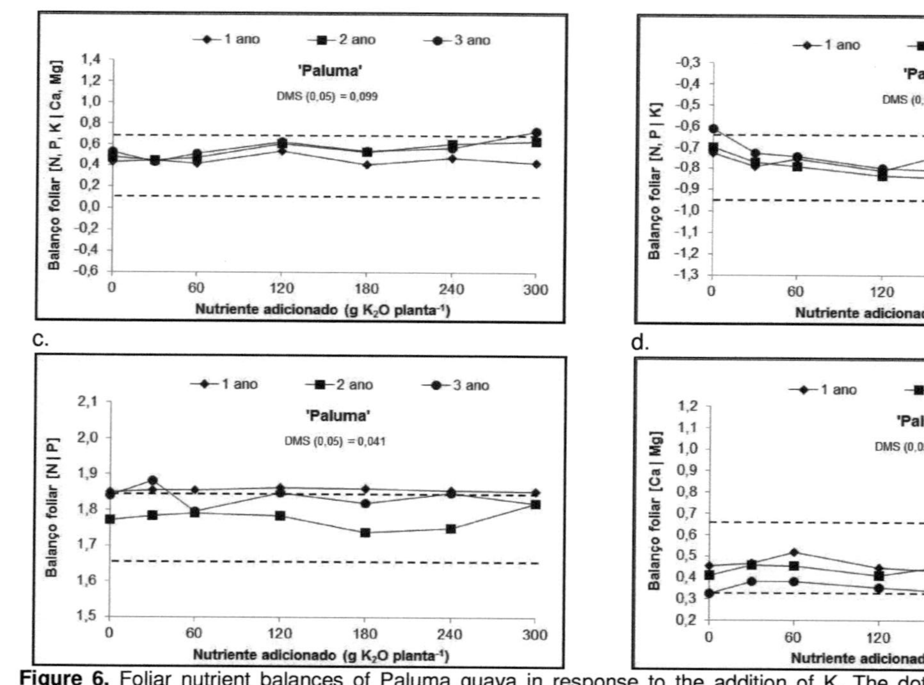

Figure 6. Foliar nutrient balances of Paluma guava in response to the addition of K. The dotted lines indicate the range of critical balance. DMS = minimum significant difference at 0.05 probability. [†] K doses in the 1° year: zero, 30, 60, 120, 180, 240 and 300g of K2O plant-1, in the 2° and 3° years double and triple the initial doses, respectively.

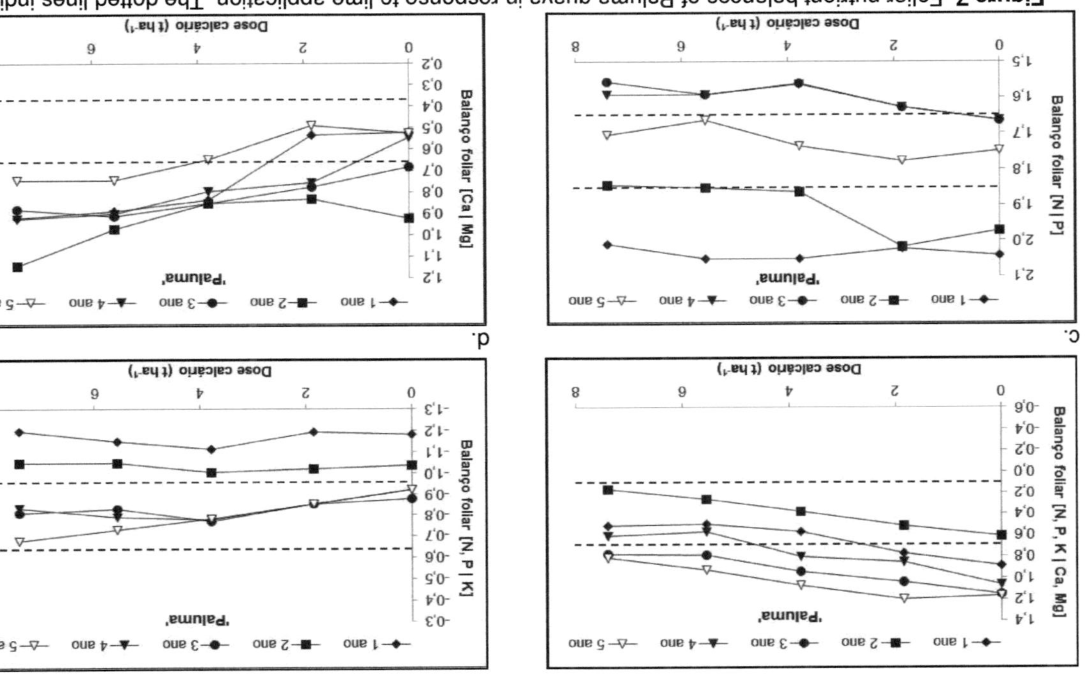

Figure 7. Foliar nutrient balances of Paluma guava in response to lime application. The dotted lines indicate the range of critical balance.

Cases of nutrient imbalance occurred during the first two years of harvest in the liming experiment (Figure 7) due, probably, to the limited root system resulting from the level of soil acidity. In fact, the expansion of the root system of newly cultivated guava plants is very sensitive to soil acidity (PRADO; NATALE, 2004). The foliar balance [N, P, K | Ca, Mg] (Figure 7a) was improved in the fifth year after liming at the highest dose, due to the increased availability of Ca and Mg. The foliar balance [N, P | K] (Figure 7b) and [N | P] (Figure 7c) were close to the adjusted foliar balance ranges reported by Natale et al. (2002b) in the second year after liming, indicating a positive effect of liming on P and K availability. The foliar balance [Ca | Mg] was above the ideal range (Figure 7d), except in the fifth year, when the liming dose did not exceed 3.8 t ha-[1] (treatment D2). Although the limestone used contained both Ca and Mg, the Ca/Mg ratio was apparently too high. Although the lime dose of 3.8 t ha-1 may be sufficient to achieve the highest level of guava yield and, adequate foliar nutrient balances, a higher dose of lime may be necessary when high doses of ammonium-containing fertilizers are applied. Natale et al. (2007) and Prado and Natale (2004) recommend a minimum limestone dose of 5.6 t ha-1 to maintain the base saturation of CEC close to 55% and pH (CaCl2) above 5.3 for at least 40 months after the application of the corrective.

Conclusion

The production of guava and the foliar nutrient balance, through a wide range of fertilization and liming regimes, allowed the definition of foliar balance ranges that tended to differ from the present ones, both at the lower and upper limits. Guava was more responsive to K and liming than to other nutrients. Liming proved to be of extreme importance in the mineral nutrition of guava, showing greater influence on the annual variation in the leaf nutrient balance, when compared to the influence of fertilization. The *ilr* proved to be a viable and reliable index for foliar nutrient balance calculations.

References

AITCHISON, J. **The statistical analysis of compositional data.** London: Chapman & Hall, 1986. 416 p.

ANJANEYULU, K.; RAGHUPATHI, H. B. Identification of yield-limiting nutrients through DRIS leaf nutrient norms and indices in guava (*Psidium guajava*). **Indian Journal of Agricultural Sciences**, New Delhi, v. 79, p. 418-421, 2009.

ARORA, J. S.; SINGH, J. R. Effect of nitrogen, phosphorus and potassium sprays on guava (*Psidium guajava* L.). **Journal of the Japanese Society for Horticultural Science**, Kyoto, v. 39, p. 55-62, 1970.

BATAGLIA, O. C.; FURLANI, A. M. C.; TEIXEIRA, J. P. F; FURLANI, P. R.; GALLO, J. R. **Métodos de análise química de plantas.** Campinas: Agronomic Institute, 1983. 48 p.

BATES, T. E. Factors affecting critical nutrient concentrations in plants and their evaluation: a review. **Soil Science**, Baltimore, v. 112, p. 116-130, 1971.

BERGMANN, W. **ErnährungsstörungenbeiKulturpflanzen.** 2nd. ed. Stuttgart: Auflage, Gustav Fischer Verlag, 1988. 762 p.

BOULD, C. Leaf analysis as a diagnostic method and advisory aid in crop production.
Experimental Agriculture, Dundee, v. 4, p. 17-27, 1968.

EGOZCUE, J. J.; PAWLOWSKY-GLAHN, V. Compositional analysis of data in Geoenvironmental Sciences. **Boletín Instituto Geológico y Minero**, Madrid, v. 122, p. 439-452, 2011.

EGOZCUE, J. J.; PAWLOWSKY-GLAHN, V. Groups of parts and their balances in compositional data analysis. **Mathematical Geology**, New York, v.37, p. 795-828, 2005.

EMBRAPA. **Sistema Brasileiro de classificação de solos.** 2nd ed. Rio de Janeiro, Brazil: Embrapa Solos, 2006. 306 p.

HAAG, H. P.; MONTEIRO, F. A.; WAKAKURI, P. Y. Frutosdegoiaba : development and nutrient extraction. **Scientia Agricola**, Piracicaba, v. 50, p. 413-418, 1993.

IHAKA, R.; GENTLEMAN, R. R: A language for data analysis and graphics. **Journal of Computational and Graphical Statistics**, Alexandria, v. 5, p. 299-314, 1996.

ISAGI, Y.; SUGIMURA, K.; SUMIDA, A.; ITO, H. How does masting happen and synchronize?. **Journal of Theoretical Biology**, Amsterdam, v. 187, p. 231-239, 1997.

KRULL, E. Soil organic matter decomposition and turnover in a tropical Ultisol: evidence from $\delta 13C$, $\delta 15N$ and geochemistry. **Radiocarbon**, Tucson, v. 44, p. 93-112, 2002.

LAGATU, H.; MAUME, L. The variations of the sum N + P_2O_5 + K20 per 100 parts of dry material of the leaf of a cultivated plant. **Comptes Rendus de l'Académie D'Agriculture de France**, Paris, v. 21, p. 85-92, 1935.

LEVY, G. J.; van der WATT, H. V. H.; SHAINBERG, I.; du PLESSIS, H. M. Potassium-calcium and sodium-calcium exchange on kaolinite and kaolinitic soils. **Soil Science Society of America Journal**, Madison, v. 52, p. 1259-1264, 1988.

MAIA, J. L. T.; BASSOI, L. H.; SILVA, D. J.; LIMA, M. A. C. D.; ASSIS, J. S. D.; MORAIS, P. L. D. D. Assessment on nutrient levels in the aerial biomass of irrigated guava in São Francisco valley, Brazil. **Revista Brasileira Fruticultura**, Jaboticabal, v. 29, p. 705-709, 2007.

MALAVOLTA, E. **Manual de nutrição de plantas.** São Paulo, Brazil: Ceres, 2006. 638 p.

MONSELISE, S. P.; GOLDSCHMIDT, E. E. Alternate bearing in fruit trees. **Horticultural Review**, New York, v. 4, p. 128-173, 1982.

NATALE, W. **Diagnose da nutrição nitrogenada e potássica em duas cultivares de goiabeira (*Psidium guajava* L.), durante três anos.** 1993. 150 f. Thesis (PhD in Soil and Plant Nutrition) - Escola Superior de Agricultura "Luiz de Queiroz", Universidade de São Paulo, Piracicaba, 1993.

NATALE, W. **Resposta da goiabeira à adubação fosfatada.** 1999. 132 p. Thesis - Faculdade de Ciências Agrárias e Veterinárias - Unesp, Jaboticabal, 1999.

NATALE, W.; COUTINHO, E. L. M.; BOARETTO, A. E. PEREIRA, F. M. Effect of potassium fertilization in 'Rica' guava cultivation. **Indian Journal of Agricultural Sciences**, New Delhi, v. 66, p. 201-207, 1996.

NATALE, W.; COUTINHO, E. L. M.; BOARETTO, A. E., BANZATTO, D. A. Phosphorus foliar fertilization in guava trees. In: TAGLIAVINI, M. (Ed.). Proceedings ISHS on Foliar Nutrition, **Acta Horticulturae**, Bologna, v. 594, p. 171-177, 2002a.

NATALE, W.; COUTINHO, E. L. M.; BOARETTO, A. E.; BANZATTO, D. A. Nutrient foliar content for high productivity cultivars of guava in Brazil. In: TAGLIAVINI, M. (Ed.). Proceedings ISHS on Foliar Nutrition, **Acta Horticulturae**, Bologna, v. 594, p. 383- 386, 2002b.

NATALE, W.; JUNIOR, J. M.; BOARETTO, A. E.; SIMÕES, F. L. Mineralogy and forms of potassium in red yellow latosol of a guava (*Psidium guajava*) tree orchard. **Indian Journal of Agricultural Sciences**, New Delhi, v. 71, p. 166-170, 2001.

NATALE, W.; PRADO, R. M.; MÔRO, F. V. Alterações anatômicas induzidas pelo calcio na parede celular de frutos de goiabeira. **Pesquisa Agropecuária Brasileira**, Brasília, v. 40, p. 1239-1242, 2005.

NATALE, W.; PRADO, R. M.; ROZANE, D. E.; ROMUALDO, L. M. Efeitos da calagem
on soil fertility, nutrition and productivity of guava. **Revista Brasileira de Ciência do Solo**, Viçosa, MG, v. 31, p. 1475-1485, 2007.

OSMAN, S. M.; ABD EL-RAHMAN, A. E. M. Effect of slow release nitrogen fertilization on growth and fruiting of guava under Mid Sinai conditions. **Australian Journal of Basic and Applied Science**, Pakistan, v. 3, p. 4366-4375, 2009.

PARENT, L. E. Diagnosis of the nutrient compositional space of fruit crops. **Revista Brasileira de Fruticultura**, Jaboticabal, v. 33, p. 321-334, 2011.

PARENT, L. E.; DAFIR, M. A theoretical concept of compositional nutrient diagnosis. **Journal of the American Society for Horticultural Science**, Mount Vernon, v.117, p.239-242, 1992.

PEREIRA, F. M. 'Rica' and 'Paluma': new guava cultivars. In: CONGRESSO BRASILEIRO DE FRUTICULTURA, 7, 1983, Florianópolis. **Proceedings...** Florianópolis: Brazilian Society of Fruitculture/EMPASC, 1984. p. 524-528.

PRADO, R. M.; NATALE, W. Calagem na nutrição de calcio e no desenvolvimento do sistema radicular da goiabeira. **Pesquisa Agropecuária Brasileira**, Brasília, v. 39, p. 1007-1012, 2004.

PRADO, R. M.; NATALE, W. Effect of liming on the mineral nutrition and yield of growing guava trees in a typic Hapludox soil. **Communications in Soil Science and Plant Analysis**, Philadelphia, v. 39, p. 2191-2204, 2008.

RAIJ, B. van; QUAGGIO, J. A.; CANTARELLA, H.; FERREIRA, M. E.; LOPES, A. S.;
BATAGLIA, O. C. **Análise química do solo para fins de fertilidade.** Campinas: Cargill Foundation, 1987. 170 p.

SANTOS, R. R.; QUAGGIO, J. A. Guava. In: RAIJ, B.van.; CANTARELLA, H.; QUAGGIO, J. A.; FURLANI, A. M. C. (Ed.). **Recomendações de adubação e calagem para o Estado de São Paulo.** Campinas: Agronomic Institute, 1996. p. 143.

SAS INSTITUTE. **SAS/STAT**: user's Guide. Version 9.2. Cary: SAS Institute, 2009. 7869 p.

SMITH, G. S.; ASHER, G. J.; CLARK, C. J. **Kiwifruit nutrition:** diagnosis of nutritional disorders. Southern Horticulture, 1985. 56 p.

TOLOSANA-DELGADO, R.; van den BOOGART, K. G. Linear models with compositions in R. In: PAWLOWSKY-GLAHN, V.; BUCCIANTI, A. (Eds.). **Compositional data analysis:** theory and applications. New York: John Wiley and Sons, 2011. p. 356-371.

van den BOOGART, K. G.; TOLOSANA-DELGADO, R. "Compositions": a unified R package to analyze compositional data. **Computers & Geosciences, [S.I.]**, v. 34, p. 320-338, 2008.

WALWORTH, J. L.; SUMNER, M. E. Foliar diagnosis: a review. **Advances in Plant Nutrition**, New York, v. 539, p. 193-240, 1988.

ZHANG, M.; HE, Z. Long-term changes in organic carbon and nutrients of an Ultisol under rice cropping in southeast China. **Geoderma**, Amsterdam, v. 118, p. 167-179, 2004.

Buy your books fast and straightforward online - at one of world's fastest growing online book stores! Environmentally sound due to Print-on-Demand technologies.

Buy your books online at
www.morebooks.shop

Kaufen Sie Ihre Bücher schnell und unkompliziert online – auf einer der am schnellsten wachsenden Buchhandelsplattformen weltweit! Dank Print-On-Demand umwelt- und ressourcenschonend produziert.

Bücher schneller online kaufen
www.morebooks.shop

KS OmniScriptum Publishing
Brivibas gatve 197
LV-1039 Riga, Latvia
Telefax: +371 686 204 55

info@omniscriptum.com
www.omniscriptum.com

Printed by Books on Demand GmbH, Norderstedt / Germany